TOWNS AGAINST TRAFFIC

Towns against Traffic

Stephen Plowden

ANDRE DEUTSCH

First published 1972 by
André Deutsch Limited
105 Great Russell Street London WC1

Printed in Great Britain by
Butler & Tanner Limited
The Selwood Printing Works Frome and London

ISBN 0 233 96312 X

Contents

List of Figures

ACKNOWLEDGEMENTS

The ideas in this book have been formulated over several years in the course of reading, discussion and work on projects. I have been influenced by the work and ideas of a great many people in ways that it is not easy to identify. I would particularly like to mention what I have learnt from Michael Thomson, whose writings are referred to several times in the book, and from the work in urban transport of my colleagues in the French companies of the Metra Group. But neither Mr Thomson nor my colleagues would necessarily endorse all the views expressed. On a more practical level, I am very grateful to my employers, Metra Consulting Group Limited, for allowing me time to write the book and for providing the secretarial and research facilities which made it possible to write it within the time. I am indebted to the London Motorway Action Group and the London Amenity and Transport Association for permission to reproduce certain sentences and paragraphs from *Transport Strategy in London*, and also to the Controller of Her Majesty's Stationery Office, the Greater London Council and the editor of *Traffic Engineering and Control* for permission to quote extracts from various documents of which they hold the copyright; the sources of all extracts quoted are given in the appropriate notes. I am also grateful to the Greater London Council for permission to reproduce the map shown as Figure 11, which is a simplified version of a map which previously appeared as Figure 1 of *Covent Garden's Moving*. This map and other maps are ultimately derived from Ordnance Survey material and are published with the sanction of the Controller of Her Majesty's Stationery Office. All the maps should, however, be regarded only as illustrations of the text; they are not intended to be accurate in detail.

Introduction

Most people do and must live in towns; the efficient organisation of life in towns is therefore one of the more important issues facing humanity. The problems concerning transport that town life raises are particularly stubborn in character and far-reaching in effect. It is a matter of common observation that this side of town life is in an unsatisfactory state at present; the main argument of this book is that it will continue to be so until a radical change is made in society's approach to the problems.

The traditional approach is first described and appraised in general terms, then the argument is illustrated in detail by reference to two particular towns, Oxford and London. It is shown that there is no solution to their problems within the framework of traditional principles, but that once these are set aside promising possibilities immediately begin to emerge. The argument then reverts to a more general plane. An account is given of the kind of measures which will have to be adopted if the problems are to be satisfactorily solved, but which have been insufficiently studied so far because they have been widely regarded as unhelpful or inadmissible. A method of study is described which can help to choose the particular measures to be adopted in any given town.

The reasons that Oxford and London were chosen as examples are that both cities are familiar to me from previous work, their problems are unusually well documented and between them they illustrate all the points that this book is intended to make. But many other examples would have served; in one form or another, these problems arise in towns and cities throughout the developed world.

The problems dealt with in this book are a proper matter of concern for all citizens and no previous technical knowledge is assumed on the reader's part. Some technicalities are introduced and explained in Chapter 3; even these, though they will be helpful to any reader who wants to go into detail, are not essential for an understanding of the main lines of the argument. The fact that the book is intended primarily for the layman does not mean, however, that it is merely a popularisation of ideas common among experts or that any central point has been omitted or treated less fully than it deserves. To write a popularisation of ideas generally accepted by experts would not be possible, if only because there is at present very little agreement among them. Nor is this surprising, since the most important questions that arise are not matters of expertise. The problem is to establish values and priorities and to decide what lines of attack are legitimate and what are not; in other words, to define the framework and terms of reference within which experts should work. This is a problem for society, not for the experts themselves. No amount of development in expertise could solve it; even the direction in which expertise should develop cannot be known until it is solved.

1. The Traditional Approach

There are two sides to the problem of urban transport. Firstly, it is difficult to move about, by any means of transport; and secondly, traffic makes towns disagreeable to live in. These problems are not new; neither are attempts to deal with them. Julius Caesar prohibited the movement of carts in Rome in the hours of daylight; his ban was extended by Claudius to the other municipalities of Italy and by Marcus Aurelius to every city in the Empire. Whatever benefits this policy may have brought to pedestrians and litter-borne travellers, the incessant noise of traffic at night made sleep impossible, at least according to Juvenal's satires.[1] In 1520, civic planners in Paris warned that 'the number of carts is growing from year to year; in a few years, we will be paralysed'.[2] In London, too, traffic has been a source of complaint and concern for several centuries.

But, although it is as well to be reminded that this is a long-standing problem, and certainly does not stem from the invention of the motor vehicle, it is also true that the motor vehicle has altered the nature of the problem and vastly increased its scale, besides opening up new ways of coping with it. So for the purpose of this book, it is legitimate to skip two millennia of human history and to start a mere generation ago. It was during the Second World War and in the immediate post-war years that serious attempts were first made to set out principles to deal with the problems caused by the use of motor vehicles in towns. These are the principles that still dominate practice, and even thinking, today.

The most important source for traditional ideas is the manual *Design and Layout of Roads in Built-up Areas* issued by the Ministry of War Transport in 1946. The main principle guiding the provision of new roads was very simple: it was to predict the growth of traffic, particularly peak-hour traffic, and to provide for it.[3] The manual gave the traffic capacities of different types and widths of road in urban areas.[4] These could be used to check the capacity of the existing roads in order to see whether they would be adequate to cope with the predicted traffic, and, if not, to ascertain what road widening or new building would be required to make them so.

This may be called a policy of indiscriminate provision of road space. The only condition necessary to justify new roads was that they should be used to a stated degree. The policy governing the use of the road network was also indiscriminate: any motor vehicle could use the roads at any time. It was recognised that public service vehicles (buses, trolley buses and trams) were important because most people were completely dependent on them, and because they occupied less road space per passenger than did cars, but there was no suggestion that they deserved any priority on those grounds.[5] Nor was the possibility raised that if some control were instituted over the use of the roads, it might lead to a reduction in the need to provide new roads. The traffic engineer's job was to provide for the safe, expeditious and orderly movement of the actual or expected volumes of traffic. The volumes of traffic themselves were not to be questioned: they were the starting point of the problem.

The principles governing both the use and provision of roads can be summarised in a single sentence, 'All traffic demands should be met.' This did not, however, imply that any priority should be given to motor traffic over other travellers or other activities. Exactly the same principles applied to cyclists and pedestrians[6] who were to be provided for by means of separate cycle tracks and pavements, the size of which was dependent upon the expected flow. The need to consider amenity and to create precincts free from heavy traffic flows was also stressed.[7]

These very general principles are too imprecise to offer much guide to the engineer or town planner who has to decide in detail what should be done, and they were supplemented by design principles. The key to good design was the segregation of traffic. It has already been said that the segregation of pedestrians and cyclists from motor vehicles was recommended; similarly, moving vehicles were to be separated from parked vehicles and different types of moving vehicles from each other. In particular, vehicles moving in different directions should be separated; through traffic from local traffic; and slow from fast-moving vehicles.

The most important step in achieving segregation was the recognition of different classes of road. Sir Alker Tripp's book *Town Planning and Road Traffic*, published in 1942, recognised three classes of road, arterial, sub-arterial and local.* The function of arterial roads was to provide 'for long distance movement through the country, and for the heavy main traffic-flows in towns'. On such roads 'every other consideration must give way to the single aim of free and rapid flow'. That meant that on arterial roads there must be no pedestrians, no standing vehicles, and as few junctions as possible. In order to minimise the number of junctions, the arterial roads should link only with each other and with sub-arterial roads, not directly to local roads.[8]

The sub-arterial roads were an intermediate class linking up the arterial roads with local roads and designed, as far as possible, on the lines of arterial roads. Local roads were roads for residence, business and shopping. They were to be designed in a way which would discourage through traffic – that is all traffic which had no business in the locality.

In existing towns, where roads were of a nondescript character, each serving several purposes, the problem was to decide which role each road was to play and then to take appropriate measures to see that that role was safeguarded. In particular, local roads were to be safeguarded by selecting the ones which should give access to sub-arterial roads, and sealing off the others. In this way, pockets would be created within the network of sub-arterial roads, 'each of which will consist of a little local system of minor roads, devoted to industrial, business, shopping or residential purposes'. The name 'precinct' was given to these pockets.[9]

Some guidance was also given as to the general shape which a town's road system should assume. The choice was seen to lie between several geometrical patterns. The manual paid most attention to a ring and radial pattern, if only because most

---

* Sir Alker Tripp's book first set out the design principles which later become incorporated into the manual and inspired many designs for towns. In some ways his reasoning is clearer than that of the manual, and in particular his classification of roads is simpler and more satisfactory, so it is easier to follow his account at this point.

English towns had developed in that way.[10] Sir Alker Tripp positively recommended it, because it involved fewer intersections than alternative designs.[11] An important function of the inner ring was to keep through traffic out of the central area. Since any ring road was likely to be longer in distance than roads through the centre, it was necessary that it should be built to the highest standard in order to make it fast enough to compensate for the extra length; otherwise the through traffic would not take it. At the same time, care was to be taken to ensure that the design of the central area network was such as to discourage through traffic.

The main point of these design ideas, particularly the idea of segregation, was to achieve road safety. But design was also important in achieving greater capacity on the arterial roads and hence on the network as a whole, and also in reducing the impact of traffic on the town's environment. There was no conflict between these objectives; a good design would simultaneously serve them all.

A critique of these principles

The policy of providing for all traffic demands implies an open-ended commitment on costs. Given this principle, it is no longer relevant to ask what benefit is achieved by satisfying the demands, or whether the cost is excessive. The question of cost can only arise when there are alternative ways of providing the amount of road space that the principle indicates, and it was only in this context that costs were discussed in the manual. The aim, then, was to steer between extravagance, on the one hand, and false economy, on the other; and of the two false economy seems to have been regarded as the greater danger.[12]

Clearly, no principle which implies an open-ended commitment to expenditure in one sector of the economy can be a sound economic principle, since it may lead to the complete neglect of other sectors – schools or housing, for example – for the sake of quite trivial benefits. The principle 'all traffic demands should be met' is economically naïve. But this would be a theoretical criticism only, if in fact the sums involved in following the principle were not very great; that is, if the increase in traffic to be predicted and provided for was relatively small. This is what was expected at the time. It was thought that the traffic volumes to be provided for over the following twenty years would in general be about double those which obtained before the war[13] – and there must have been many roads, at least outside the centres of large cities, which could have carried twice their pre-war peak-hour flows without being widened. Sir Alker Tripp suggested that proper use of the existing road space, through the implementation of the design principles he suggested and especially through the control of on-street parking, would suffice to supply at least the present needs of many towns.[14] But experience has not borne this out; traffic has everywhere increased much more than expected. The fundamental weakness of traditional ideas was that the phenomenon of traffic growth was not properly understood.

Growth in traffic was attributed to 'demand' factors outside the planner's control – population, vehicle ownership, incomes and so on, and these factors were themselves underestimated. For example, the number of vehicles registered in 1966 was more than four times the number registered in 1939, not twice the number as had been expected. But even more serious was the failure to realise the connection between road-building and traffic volumes: the fact that providing more road space itself generates more traffic. It is not just that new roads divert traffic from less attractive old roads – this effect was well understood – but that the total number of miles performed on a road network is related to the capacity of the network, the larger the town, the closer being the relationship. Very broadly speaking, the amount of traffic is governed by what is regarded as a tolerable level of congestion.* If the capacity of the road network is increased, whether by road construction or by traffic management measures, the mileage will increase until the same conditions obtain. If the capacity of the road network is not increased, the mileage performed will stabilise, and if the capacity is reduced, the mileage will be reduced correspondingly.

* It is the marginal user – the person for whom the difficulties of making the journey just about balance with the benefit derived from making it – who finds the existing level of congestion just about tolerable. This does not mean that the level of congestion is or ought to be acceptable to other road users or to the community as a whole.

It is not clear just how this process works, but its existence has long been recognised, especially in America, where road building since the war has been on a greater scale than in any other country, but where congestion in large urban areas has not been cured. An American author wrote in 1963:[15]

Traffic engineering has performed creditably and has achieved much of what was expected of it. The technical skills acquired by the traffic engineer through practice and research have aided measurably in expediting traffic movement and in reducing driving hazards. (The death rate from motor vehicles in the United States is now one-third the rate in 1930; while the traffic engineer cannot take all the credit for this reduction, he is entitled to some of it.)

But post-war urban growth has brought with it traffic and transportation problems of unprecedented complexity and difficulty. Our efforts to find satisfactory answers to these problems have not been wholly successful, mainly, I think, because we have not correctly assessed them. We have assumed that where there was an obvious lack in street and highway capacity or parking, we could solve the problem simply by providing what was needed, either by making better use of what we had or by building new facilities. However, we have been learning during the past ten or fifteen years that this piecemeal approach is not adequate to cope with the increasingly difficult transportation problems which seem to arise from the very successes of traffic and highway engineering.

As traffic engineering measures and new streets and highways have added capacity to road systems, more motor vehicles have been put into use almost immediately. New roads and more motor vehicle miles of travel have gone hand in hand in the United States. . . .

In Britain, too, there is evidence that flows increase with road capacity in such a way that speeds remain virtually constant. This is not completely conclusive, however, since it could be argued that the increases in capacity were just sufficient to accommodate the flows that would have come about anyway. This appears to be the conclusion drawn from a study of traffic conditions on main roads in eight English provincial towns in 1965 and 1967.[16]

Between 1963 and 1967, despite a 15% increase in the mean off-peak flow, the speed fell by only one kilometre per hour (3%). A similar decrease in speed was noted in the central area of the towns during this period but in this case the increase in the mean flow was smaller (8%). In the evening peak period the changes in speed and flow were slight with the 1967 mean speed in the central area showing a small increase. A general decrease has taken place in street parking, particularly in the central areas and this, together with the introduction of traffic management schemes, helped to keep speeds relatively constant despite rising flows in the four-year period.

The most famous example of the opposite effect – the contraction in the number of vehicle miles performed when the capacity of the road network is reduced – is Washington Square Park in New York. This case has been described at some length by Jane Jacobs, from whose account the following summary is taken.[17] The local residents were threatened with a plan to build a major highway through this park to replace an existing main road, the capacity of which was not thought adequate to cope with proposed developments in the surrounding area. Instead they managed, after a political battle, to have the existing road closed, first on a trial basis and then permanently. The traffic commissioner, in opposing this scheme, had forecast immediate and very severe increases in the number of vehicles in the nearby streets, to the extent that the residents themselves would be obliged to ask for the park road to be re-opened. In fact, none of the surrounding roads experienced an increase in traffic, and most experienced some decrease. Nor was there any sign that the traffic had chosen more distant alternative routes in other parts of the city; it had simply disappeared.

Similar experiences have been claimed in Europe when shopping streets have been converted to exclusive pedestrian use, although it is not clear how good or extensive the before and after counts were,[18] but to demonstrate the existence of the relationship between road capacity and traffic volumes it is not necessary to appeal to the results of the rather small number of examples that have been studied scientifically. The overwhelming evidence is the everyday experience of large towns and cities all over the developed world. None of them has experienced a complete breakdown of traffic,* though in all of them congestion

* There have been sensational traffic jams over wide areas on particular days. These are presumably attributable to a number of breakdowns, of vehicles, traffic lights, etc, occurring simultaneously on different critical parts of the network after a certain volume of traffic has been committed to it. They have sometimes been taken as heralding the imminent and permanent seizure of the road network, but such prophecies have not been borne out.

is a problem and source of complaint. Either there is some automatic mechanism whereby the volume of traffic adjusts itself to suit the capacity of the road system, or every town has grown up in such a way that the capacity of the streets is just sufficient to cope, albeit at a rather unsatisfactory standard, with an independently determined demand. The latter hypothesis implies such amazing coincidences that it cannot be taken seriously.

The mechanism which produces the adjustment is a very subtle one, which it may never be possible to describe in terms of quantified theory showing all the interactions, but it may be helpful merely to list the ways in which new roads lead to greater mileage. The first order effects are that drivers may perform the same journeys as before but by longer (though faster) routes, or may make different, longer journeys or may make more journeys. Simultaneously, new roads will encourage a shift from public to private transport. The chief second order effect is that the shift from public to private transport will lead to a decline in the service or an increase in fares on public transport, which will further encourage the shift.

New roads will also encourage people to live further from their work or from other places of activity, which means that their journeys become longer and are less easily catered for by public transport. This kind of dispersal can take place to some extent within an existing stock of buildings, but is very much encouraged by the tendency of development to take place either along the lines of new roads or at least in places convenient to them. As a result of all these tendencies, the attraction of owning a car, or the deprivation of being without one, is increased, so that car ownership is stimulated; in particular the tendency for one household to own two or more cars. Finally, increased car ownership itself leads to increased travel.*

Clearly, there must be some limit to the process of traffic generation. No one wants to drive around all day simply because there are enough roads to make it possible to do so. There are small and medium-sized towns whose traffic problems are confined to two fairly short peaks, and even in the centres of large cities congested conditions rarely obtain for twenty-four hours

---

* For a much more detailed description of the processes involved, the reader is referred to *Motorways in London.*[19]

a day. But even though it might be possible to construct roads on such a scale that all the traffic that would then arise could be accommodated to some pre-determined uncongested or free-flow standard, the sums of money required are likely to be very large. In some towns, it might be possible to raise the money, but the question of how to justify the expenditure would become a real and urgent, not simply an academic, question. In other towns, the money required would far exceed what could possibly be made available: for practical purposes there would be no limit to traffic generation. In the first case, the rule 'predict and provide' would be a very bad guide, since to follow it is almost certain to be very wasteful. In the second case, the rule ceases to have any meaning, since the volume of traffic to be predicted depends entirely on the scale of provision made. In order to predict, one has to know the very thing which the prediction is supposed to help one decide.

The attempt to follow traditional policies has had other unfortunate consequences. Although in theory, as we have seen, there is no presumption that any one activity should have priority over any other, in practice the various objectives are irreconcilable and some sort of system of priorities has to emerge. In the next chapter it is argued that the ordering of priorities which we have tacitly adopted, since it is implicit in the practical arrangements which we have allowed to continue, cannot be defended once it has been made explicit. Another serious consequence is that these policies tend to reinforce the existing pattern of development, whether or not that pattern is convenient. In most English towns travel is directed on the town centre, because that is where the main travel generators are and because the road network is itself of a radial pattern focused on the centre. The radial roads become particularly congested, congestion is seen as evidence of demand, and the roads are enlarged or new roads serving the same function are built. The accessibility of the centre relative to that of other areas is thereby enhanced, which attracts more shops or businesses of a kind which generate many journeys, and so the process continues. The resulting pattern may be one that gives rise to long and awkward journeys. The circle is vicious in another way too, in that it tends to destroy many of the more interesting parts of the town. The construction of new roads itself often involves destruction of buildings. But in addition to that, the increased accessibility of central areas leads to higher rents and therefore to a demand to replace old buildings by larger, more modern buildings which can justify the rents. This is the story of many English towns.

It is unlikely that anyone would now consciously defend traditional policy principles: the difficulty is, as we shall see, that they are nevertheless presupposed by the methods of study still in use. Traditional design principles have survived much better. The Buchanan Report, *Traffic in Towns*, attacked the geometrical kind of thinking which underlay traditional ideas about network design (see Chapter 3) but helped to popularise and establish the more detailed principles of the design of roads and precincts that had been put forward before, particularly by Sir Alker Tripp. These ideas, and particularly the principle that heavy flows of motor vehicles should be kept physically separate from other activities, and mutually conflicting kinds of traffic from each other, are obviously valuable and deserve to survive. But there is one point on which they have introduced some confusion in thought: the idea that traffic of different origins and destinations should be segregated, in particular that through traffic should be separated from local or stopping traffic and long distance from short. This idea is now firmly embodied in official thinking, and has an immense influence on the design of road proposals, being sometimes regarded as the key both to environmental and traffic problems. Nevertheless, it is very puzzling.

For example, in 1966 the Ministry of Transport issued a publication *Roads in Urban Areas* which replaced the 1946 manual. In this publication it is stated that:

Traffic segregation should be the keynote of modern road design and should be arranged to reduce the risk of conflict between one vehicle and another and between motor vehicles and slower-moving and more vulnerable road users such as pedestrians and pedal cyclists. Some examples of the application of segregation to the urban road system are outlined in Table 1.2.

The first example given in the table is segregation of traffic by origin and destination, which seems to have nothing to do with the conflicts mentioned.[20]

But this idea has never been without its critics. For example, a speaker at a conference of civil engineers in 1956 was reported thus:

The popular idea in the past of by-passes, ring roads, etc was that they were to divert through traffic from urban or residential areas as if through traffic was potentially more dangerous than local traffic. Probably one of the most dangerous vehicles, in proportion to its mileage, was the electric milk-delivery van, at any rate as far as young children were concerned, and that could hardly be called a through vehicle.[21]

From the point of view of the traffic, there is no particular virtue in separating long- and short-distance traffic, or traffic bound to one destination from that bound to another. Whether a driver is making a long journey or a short one, he wants to travel smoothly and quickly; it is of no concern to him that the other vehicles with which he shares the road and which may impede him are making journeys longer or shorter than his own. From the point of view of the environment, what matters is that the total volume of vehicles, and perhaps also the number of vehicles of particular types, should be limited; and that the traffic should move in a way which does not threaten the safety and convenience of pedestrians or the general harmony of the area. So long as these conditions are met, it does not matter whether the traffic is going right through the area or stopping in it; if they are not met, it is no consolation that all the traffic within the area has business there.

There are, of course, places where the through traffic constitutes such a large proportion of the total traffic that to divert it in some way would be a sufficient means of achieving reasonable environmental conditions. A town built on a major national road would be a prime example. Sir Alker Tripp seems to have had such instances particularly in mind, and this presumably led him to think of the general problem in terms of eliminating through traffic. His thought was probably also influenced by the means and control at his disposal, since it is in this connection that the distinction between through traffic and local or stopping traffic really does become important. Some methods of control, such as road pricing, operate on both kinds of traffic. Parking control acts only on stopping traffic. But there was little thought of either of these methods when Sir Alker Tripp was writing. As has been seen, he thought mostly in terms of the physical design of the network, sealing off particular roads. The effect of such measures is that it is impossible, or at least never advantageous, for traffic to filter through an area, but traffic destined for the area will not be deterred.

Another reason why Sir Alker Tripp thought it desirable to separate through from local traffic is that he imagined that the through traffic would be going fast, and indeed should be allowed to travel fast if full advantage were to be taken of motor vehicles. This is reasonable in relation to traffic on a major national route which traverses a town, but is much less convincing in relation to journeys lying entirely within a town. High travelling speeds are not very important for such journeys; to be able to start one's journey at any time and to penetrate very close to one's final destination are usually much more significant advantages.

From the environmental point of view, the example of main road traffic running through a small town tends to mislead.* If a by-pass is built round such a town, the through traffic can be removed from it altogether. But in the case of a large town the through traffic removed from a precinct must still traverse the town, either on an existing road, or on a specially built arterial. Almost all existing roads serve purposes other than traffic purposes, and the environmental loss of imposing much heavier traffic loads on them is not diminished simply by christening them arterial roads. To build large new roads in towns is damaging in itself, and opens up new areas to the noise, fumes and visual intrusion of traffic, even though physical separation, and hence safety, is achieved. These environmental losses must be set against the environmental gains of establishing precincts. Whether or not the gains outweigh the losses depends on how much reconstruction is involved, which in turn depends on how much traffic it is intended to accommodate.

* It is not clear to what extent Sir Alker Tripp himself was misled by this example, but some of his contemporaries were, particularly Abercombie, as will be seen in Chapter 5, and it has tended to confuse thought ever since.

2. The Case for Controls

If road building by itself will not solve our problems, there are only two alternatives. We must either permit the situation to continue much as it is, or even to deteriorate further, or we must find some way of making better use of roads. The prospect of an indefinite continuation of the present situation is sufficiently disagreeable to make a search for better ways of using the roads well worthwhile. A detailed discussion of the possibilities will be found in Chapter 9, but they all have one thing in common. They all imply some departure from the principle of the unrestricted, or indiscriminate, use of the roads; the idea that any vehicle may use any road at any time. But any suggestions that contravene this principle tend to arouse strong resistance, among some influential people at least, and are accepted, if at all, only with the greatest reluctance. It is largely because of this feeling that so little has been done to explore the possibilities. The feeling is understandable but completely misplaced and nothing is more important than to dispel it as thoroughly and as early as possible.

A number of quotations will illustrate the attitude. The first one dates from 1927; no one would use quite this language today, but the attitude survives.

But shall we ever stand such a denial of individual liberty [as a policy of restricting the use of private cars in towns would represent]? If I am right in my opinion that the right to use the road, that wonderful emblem of liberty, is deeply engrained in our history and character, such action will meet with the most stubborn opposition. More street space and more road space will have to be provided whatever be the plan for it or the cost of it.[1]

Distasteful though we find the whole idea, we think that some deliberate limitation of the volume of motor traffic in our cities is quite unavoidable. The need for it simply cannot be escaped.[2]

Some people would like to push us into a frame of mind in which it is considered anti-social to own a car; selfish to drive one; and positively sinful to take it into a built-up area. Of course traffic in towns creates a problem. . . . My approach to this problem is not to restrict, to hamper or to confine the motorist. Instead we must learn to *cope* with the motor car and to care for the motorist. . . .[3]

Two feelings are revealed in these quotations. Any control or intervention by the authorities is seen as a restrictive act – one more encroachment by the state on individual liberty. Provision is seen as a positive, active policy, in contrast with restraint which is negative and defeatist.

This attitude presupposes that provision is in itself a possible policy, whereas, as has been seen, it could not produce the desired result. This misunderstanding produces others. The most important is the idea that we have a choice between a policy which involves some restrictions and a policy which does not. There is no such choice. All policies involve some restrictions, including the present policy. The present freedom to use a motor vehicle, or to move in other ways just how and when one pleases, turns out to be only a legal fiction. In practice travellers are prevented from moving about as they wish, not indeed by rules imposed from outside, but by other travellers. The choice is only between the restrictions inherent in the workings of the present system and rules consciously worked out in accord with some idea of the general advantage.

The fact that the restrictions inherent in the present situation do not take the form of rules and regulations makes it difficult to recognise them as restrictions at all. In addition, the present system has been with us so long that we tend to regard it as the natural state of affairs, and anything else as an aberration. We accept the priorities implicit in it without question, without perhaps even realising that there are any. Nor would there be any if the system worked according to the theory, with provision made for all travellers and all activities. But because such provision is impossible, conflicts arise which are settled according to a very questionable set of priorities.

The first conflict, which is too familiar to need emphasis, is between vehicles and the environment. The rule that 'any vehicle may use any road at any time' ensures that vehicles will always win this conflict, which tends to spread from the main roads to quieter side roads, as drivers increasingly seek quicker and less congested routes. The only protection that the residents or others affected have is that the design of the network may be such that drivers do not want to use those streets because they would gain no advantage from doing so.

The second conflict is between travellers by motor vehicle and pedestrians and cyclists. Other things being equal, one should presumably start with the principle that all travellers have equal rights, regardless of the means by which they choose to travel. But since travellers by motor vehicle are better armed and better protected than pedestrians and cyclists they tend to take priority whenever any conflict arises. One aim of policy should be to correct this bias. Moreover, from the general social point of view, other things are not at all equal. Pedestrians and cyclists are much cheaper to accommodate than motor vehicles and do no environmental harm. This is a strong reason for giving them not merely equal, but preferential, treatment.

The rule 'any vehicle may use any road at any time' means that all motor vehicles are treated equally, in particular that buses and cars are treated equally. To start with the principle that all travellers should be treated equally would imply, in congested conditions, giving priority to buses because they carry more passengers. In fact buses tend to lose more than cars under congestion. They have no opportunity to seek better, less congested roads, and delays to particular buses on particular sections of the route have repercussions on the performance and reliability of the whole bus system.

The principle that all travellers should be treated equally is itself only a starting point; some account should be taken of the urgency of different travellers' journeys and of their particular travelling requirements. But the present system is extremely unselective. No priority is given to travellers who need to make their journeys to a particular place by a particular mode at a particular time over those who could travel elsewhere or by other means or at another time.

These unfortunate tendencies, which are the natural results of congestion, have been reinforced by the traffic management measures which have been traditionally used to help cope with the problems. The aim of these measures has been to increase the speed and flow of moving vehicles. Amenity suffers, since drivers have been encouraged to find alternative routes, including roads which ought not to carry heavy flows of traffic. Buses and lorries gain, as one element in the traffic flow, from increased speeds, but these gains may well be offset by longer and less convenient routes and by difficulties in stopping and unloading. Pedestrians sometimes gain, as measures intended to increase flows and speeds can sometimes, incidentally and fortuitously, work to their advantage too. Cyclists always lose.

The techniques of traffic management do not have to be used in pursuit of these particular aims. But the traffic engineer's job has traditionally been to cope with a flow of moving vehicles which for him was 'given'. So long as the problems could be dealt with at that level, there was no need to think in terms of the movements of people and goods, which are the underlying reasons for the movement of vehicles. Also there is a natural tendency in any complex problem for attention to be drawn to its most conspicuous parts. Traffic jams are headline news, but the decline of cycling is not. The eye is more naturally drawn to a long line of stationary vehicles than to a group of people waiting patiently at a bus stop or trying to cross the road. People who do not travel are the least conspicuous of all. There may be many people, especially the old, the very young and the infirm who would like to travel but are deterred by the decline in services or by the physical difficulties involved, but no attention has been paid to their problems.

Another stimulus for traffic management measures has been the fear that unless something drastic was done, the traffic would finally grind to a halt, with immensely damaging effects for the economic life of the city. To prevent such a catastrophe some sacrifice of amenity and some discomfort and inconvenience to certain travellers or would-be travellers was clearly justified, especially when the restrictions were seen as temporary only: stop-gap measures until new roads could be provided on an adequate scale. The fears were groundless but the motive was respectable.

There is very little to be said for the way that the present system works, either in its natural form or as reinforced by deliberate acts of policy. It is wasteful, arbitrary and damaging to civilised conditions, not freedom but a free-for-all. To introduce controls should, therefore, be thought of as a means of correcting the restrictions and biases inherent in the situation, not as an additional restriction.

As for the rights of the individual, there is no new principle at stake. If one person can only exercise his freedom at the expense of others, then it is right to restrict him. Moreover in this situation it is not a simple matter of one fixed class of people losing and another fixed class gaining. The system tends to favour car users against other travellers, but it is not clear that even car users in fact gain from it. People now using cars might prefer to walk or cycle or go by bus, if only conditions were better by those means. But because all the alternatives give such an indifferent service, people choose to go by car – and hence contribute to the further worsening of the alternatives. Freedom to choose between the available options is not very interesting if all the options are poor. Controls might limit the range of options, but would improve their quality. Restraint of traffic is not restraint of movement.

A numerical example may help to illustrate the way in which a situation can develop which works to no-one's advantage, although it results from every individual's acting rationally in his own interest.

Table 1 is concerned with journeys to work to Central London in the morning peak. It shows the average journey times, for car travellers, bus travellers and both kinds of traveller together, that might be expected on different assumptions about how journeys are divided between bus and car. No great realism is claimed for these figures, which are necessarily based on very broad assumptions; they are only being used to illustrate the theoretical point. For the sake of the illustration, it is assumed that each traveller wants to minimise his journey time; in reality many other considerations will weigh with travellers.

According to these figures, the greater the proportion of travellers by car, the longer the journey time for all travellers. For any particular split between bus and car, it is better to be a traveller by car than a traveller by bus, but is better to be a traveller by bus when 10% or less of journeys are made by car than a traveller by car when 50% or more of journeys are made by car. But of course the individual traveller is never confronted with this choice. He can only choose how *he* will travel, not what the overall split will be. If 10% of travellers are going by car, then

TABLE 1

Times of commuter journeys to Central London by car and bus

% of journeys made by car	MINUTES TAKEN TO TRAVEL BY		
	Car	*Bus*	*Both modes together*
0	(32·6)	50·4	50·4
10	34·5	52·7	50·9
20	36·9	55·7	51·9
30	40·5	59·8	54·0
40	45·8	66·2	58·0
45	50·1	71·2	61·7
50	55·6	77·4	66·5
60	83·5	108·2	93·4

Note: Both sets of journey times are door-to-door times, including waiting, walking and parking times. The time allowed for parking is ten minutes.

Source: Table IV of 'Car and bus journeys to and from Central London in peak hours' by P. B. Goodwin, *Traffic and Engineering Control*, December 1969.

the average bus traveller sees a substantial advantage to him in switching to car; he hopes to reduce his journey time from 52·7 minutes to 34·5 minutes. But if enough other bus travellers make the same choice, he does not in fact manage to travel in 34·5 minutes, but only in 36·9 minutes. This is still better than travelling by bus, however, particularly since the time by bus has itself increased – the journey now takes 55·7 minutes. So more people decide to switch to car and this process can continue, long after it is in everyone's interests to stop it, until a limit is reached set by some external factor: for example, the number of people who have the use of cars, or the number of parking spaces available. The object of a control would be to prevent the situation from degenerating in this way.

This sort of situation is common in urban affairs and is one of the main justifications for having a planning system at all.

The courses of action that are open when people act as a community, accepting the disciplines that such cooperation involves, are more attractive than those open to the same people acting individually. Thomas Sharp has put the point very clearly, in another context:

Men are sick of the wretched towns they have been given to live in. It is natural and right that they should want to escape . . . they will escape to Suburbia if nothing better is offered . . . But is the flight to Suburbia a real escape, after all? Every person who goes to the suburbs seeking the edge of the countryside pushes the countryside away from somebody else – and then he in turn suffers from having it pushed away from him. The inhabitants of Suburbia continually thwart themselves and each other, and the more they strive to embrace the object of their desire the more it escapes them: the more they try to make the best of both worlds the more they make the worst.[4]

It is useful to express these points in terms of the traffic engineer's concept of demand, since this concept occurs again and again in the context of road planning and obviously exerts a profound influence, although there is very little substance in it. We have seen that traffic engineers have traditionally thought it their duty to provide in a positive way for all the demands that might be made on the roads for which they are responsible. Understandably and creditably, they dislike any suggestion that they cannot and should not attempt to do so – it sounds like falling down on the job. But in fact there is no other sector of the economy in which it would be thought right to supply all demands in a limitless and indiscriminate way, nor would the recipients wish them to be supplied if it involved the destruction of other things which they valued or the commitment of large resources for which there were better uses.

In fact such 'demands' are not in an economic sense demands at all, but only vague desires, which have no particular claim on anyone's attention. The situation is still worse if the traffic flows which can be seen on the roads are interpreted as an expression of either demands or desires. The flows represent people's behaviour and choices after they have adapted themselves to the possibilities and limitations of the situation. They may be forced into a course of action which they thoroughly dislike. If a shopper buys a certain product only because the other ones he asked for are out of stock, the manufacturer would be unwise to interpret such a purchase as showing a desire for his product. But analagous situations occur constantly in urban transport. The man who waits at a bus stop for ten minutes and finally hails a taxi is only the most obvious example. But no commercial analogy can do justice to congestion; the fact that one customer buys Brand A does not produce an immediate deterioration in the quality of Brand B.

It may be objected at this point that although the argument is logical enough, public opinion would not accept it. If this were true, it would indicate a need to educate public opinion; it would not be an argument for continuing with policies which cannot achieve the intended result. However, the available evidence does not support the contention; it suggests that the public is prepared to accept controls as part of a package deal in which certain rights would be foregone in exchange for a general improvement in travelling conditions and the environment.[5]

One more analogy may help to clarify the issue. It is sometimes said that although most Englishmen live in towns they are all countrymen at heart and would really like to live in the country. Suppose that there is some such desire, no-one would suggest that the aim of housing policy should be to satisfy it. Such a policy would not succeed, vast sums would be spent in its pursuit, the attempt would destroy the very amenity which it is trying to make available, and at the same time would fail to make the best of the opportunities that really are inherent in the situation. All the same things can be said of the attempt to base urban transport policy on the principle of the unrestricted use of the roads.

It is essential always to bear the question of controls in mind when assessing the ideas that have been put forward since the war years or the schemes that have been proposed for particular towns. Has the need for controls been realised? If it has, have controls been seen in a purely negative light as something forced upon us by the difficulties of building enough roads, or in a more positive way as an essential condition of constructing a satisfactory transport system? These questions are considered in the following chapters.

3. How Thought has Developed

A generation has gone by since the ideas that were discussed in Chapter 1 were propounded. During that time the magnitude of the problems of urban transport has become steadily more apparent, much thought has gone into their study and new techniques have been developed. The object of this chapter is to see to what extent this work has produced more satisfactory and fruitful lines of approach.

'Traffic in Towns'

In 1961 the Ministry of Transport set up a working group 'to study the long term development of roads and traffic in urban areas and their influence on the urban environment'. The report was published in 1963. It was given the title *Traffic in Towns*, although it is more generally known as the 'Buchanan Report', after the leader of the working group, Sir Colin Buchanan. The report followed the principles first worked out by Sir Alker Tripp but also developed them and introduced some new ideas. The growth in traffic since Tripp's day had shown up some of the limitations of his approach, and it is this above all that accounts for the differences introduced by Buchanan.

Buchanan, like Tripp, saw the problem primarily in terms of design. His design principles are fundamentally the same as Tripp's, although the terminology is different. Buchanan distinguished only two types of road: distributors, whose sole function was to carry moving traffic, and access roads which gave direct access to land and premises. District distributors, the equivalent of Tripp's sub-arterial roads, would form the boundaries of environmental areas, the equivalent of Tripp's precincts. These district distributors would link with local distributors within environmental areas which would in turn link to the access roads.

Environmental areas would be free of all through traffic, and within them environmental considerations would predominate over traffic considerations.

Buchanan took a much stronger view than Tripp of the inadequacies of the existing layout of towns and of the corresponding need to redesign them. Tripp did not pretend that to replan in the way he suggested would be easy, or that no new roads would be required, but nevertheless he thought that once a proper classification of roads had been made and practical steps taken to deal with the problem of on-street parking and other standing vehicles the problem would be well on the way to a solution.[1] For Buchanan, however, it was the whole inherited street pattern, and not just the existence of some nondescript streets, that constituted the problem.[2] Not surprisingly, therefore, Buchanan thought in terms of smaller environmental areas, denser road networks and much more road construction than Tripp had envisaged. He also introduced the idea of a hierarchy of distributors, the distributors in the hierarchy connecting with each other much in the same way as the branches and twigs of a tree are connected. Thus, in order to get from a local distributor within an environmental area in a large town to a national distributor, it would generally be necessary to proceed via a district distributor, a primary distributor and a regional distributor. The higher the position of the road in the hierarchy, the higher the standard it would be built to. Primary distributors – the highest class within a town – would in many cases be built to motorway standards.[3]

In one important respect Buchanan's views are actually in conflict with Tripp's. Tripp had thought in terms of an ideal geometrical shape for a town's road networks, although of course by the time that the network had been fitted to a particular town it would be unlikely to retain any exact symmetry.[4] Buchanan rejected the whole of this geometrical way of thinking. The way to design the network was first to define environmental areas and then to study the amount of traffic they generated and the links, in terms of traffic, between them and other areas. The shape of the network would emerge from such a study; it might or might not turn out to have any simple geometrical pattern.[5]

As well as the horizontal segregation of traffic from other activities implicit in the concept of distributors and environmental areas, Buchanan also suggested that vertical segregation would sometimes be appropriate. This idea had been discussed by Tripp, but very much as a second best solution to be considered when proper horizontal separation would be too difficult to achieve.[6] Buchanan took a much more positive view of the possibilities. As an extreme case, he discussed the possibility of giving up the entire ground floor to motor vehicles and building a town centre, for example, on a new building deck above the traffic. In existing towns, at least, the solution would more often be found in complex multi-level arrangements which would include both pedestrian precincts at ground level and pedestrian paths above the traffic.[7]

Buchanan's design principles are, for the most part, the same as those of Tripp; but Buchanan realised, as Tripp had not, that design alone, even if it was possible to contemplate rebuilding on a massive scale, would not solve the problem. From the outset, Buchanan had stipulated some control over the use made of the roads, in that when a distributor network had been provided traffic would be obliged to use it without choice.[8] Tripp had always thought that it would be necessary to induce traffic to use the arterial roads by making them more attractive than the other roads, and this necessity had quite an important influence on his design ideas.[9] But beyond this, the practical studies undertaken for the purpose of the report convinced Buchanan that in the long term, having regard to the expected growth in population and car ownership, there was no chance of being able to accommodate all the traffic demands that would arise. In some large towns, it might be physically impossible to do so, even if complete redevelopment could be considered, at least if the town was to remain reasonably compact. In other towns the necessity, or obvious desirability, of preserving much of the existing building would severely limit the amount of road space that could be provided, and hence the amount of traffic that could be accommodated. But even when there was no physical constraint of that sort, there was always likely to be a monetary constraint. This finding gave rise to the famous 'law' that 'within any urban area as it stands the establishment of environmental standards automatically determines the accessibility, but the latter can be increased according to the amount of money that can be spent on physical alterations'.[10]

B

If it is not possible to accommodate all traffic demands, then the question arises which traffic should be accommodated and which should not, and what means of control should be introduced to discriminate between them. Buchanan suggested that a distinction should be made between essential and optional traffic. Initially at least, Buchanan defined essential traffic as that arising from the use of vehicles for essential purposes in connection with trade, business and industry, and optional traffic as that arising from the use of cars for private pleasure and convenience. Very little attention was paid in the main part of the report to possible methods of control, but in the final chapter several methods were briefly described, of which parking control was judged to be the most important and immediately promising.[11]

In appraising the ideas of *Traffic in Towns*, it is perhaps easiest to adopt the authors' own headings of environment, accessibility and cost.

To what extent would the adoption of the principles laid down really produce a satisfactory environment? Even within the environmental areas themselves, it is proposed that there should be distributor roads, that is roads given up entirely to the purpose of traffic.[12] The Amount of traffic that was to be expected within environmental areas was not defined in numerical terms but it was stated that 'it cannot be emphasised too strongly that the environmental areas envisaged here may be busy areas in which there is a considerable amount of traffic, but there is no extraneous traffic, no drifts of traffic filtering through without business in the area'.[13]

This seems to repeat the fallacious notion already discussed in Chapter 1, that the environmental nuisance of traffic is determined in some way by whether the traffic is through or local.

However, *Traffic in Towns* suggests that environmental areas cannot become excessively encumbered with their own traffic, because the size of any area is governed by its environmental traffic capacity: if there were too much traffic, the area would then have to be sub-divided. This, however, raises the possibility that environmental areas might have to be very small, and it is indeed stated in *Traffic in Towns* that 'a neighbourhood of 10,000 people, which was the unit size postulated in the County of London Plan, would certainly require sub-division into a number of environmental areas'.[14] To this, all the same objections apply, but with much greater force, as were made in Chapter 1 in relation to Tripp's ideas. To split up a town into a number of small areas is itself an environmentally damaging act, whether it is done by converting existing roads into distributors or by building new roads. The environmental effects of the traffic alongside the distributor roads at the edge of the environmental areas cannot be disregarded, especially if the areas are small and a large proportion of the buildings within them will be on the perimeter, or close enough to it for the traffic to make its presence felt.

Similar doubts arise about the efficacy of the ideas of *Traffic in Towns* for vertical segregation, attractive though the written descriptions make those ideas appear. This point is discussed in Chapter 6, on Central London, since it is there that the idea of vertical segregation has had most influence.

The most serious conceptual weakness in the analysis of *Traffic in Towns* is in the treatment of accessibility. Accessibility is defined in the Glossary as 'the degree of freedom for vehicles to circulate and penetrate to individual destinations and to stop on arrival'. This definition immediately excludes pedestrians, and although ostensibly it covers cyclists, it is clear from the way that the term is used in the text that cyclists are in fact excluded: by 'vehicles' is really meant 'motor vehicles'. But walking and cycling account for many journeys, and it is unfortunate to define the basic terms in a way which disregards them altogether. This would be a logical point of no great importance if the illustrative suggestions of *Traffic in Towns* had in fact taken the interests of pedestrians and cyclists into account, notwithstanding the definition. Pedestrians do indeed figure largely under the heading of 'environment': the main point of an environmental area is to allow pedestrians to move about with ease and safety. But this is short-distance pedestrian movement only, walking within a small area after one has arrived there by other means. Although at one point *Traffic in Towns* mentions that walking is important for medium-distance movements, it says almost nothing about how pedestrians might move between one environmental area and another, and clearly the establishment of a network of distributors on the scale suggested would make that kind of movement very much more difficult. In the same way, it would hamper cyclists, for whose problems *Traffic in Towns* showed very little concern.*

* The theoretical discussion in *Traffic in Towns* says nothing about

But even if attention is confined only to travellers by motor vehicles, the approach of *Traffic in Towns* is thoroughly unfortunate because in fact it concentrates on the problem of cars. The emphasis given in the definition of 'accessibility' to the need for vehicles to be able to stop near their destinations itself indicates this, since to make provision for stopping is a much more troublesome problem for cars (and of course for lorries) than it is for buses or taxis. But the bias becomes much clearer in the text, in which accessibility is equated with the number of motor vehicles in use or the proportion of journeys that can be made by car, thus disregarding the possibility, which as was illustrated in Chapter 2 might be a very real one, that the interests of travellers might be best served by limiting the number of vehicles in use. *Traffic in Towns* is concerned with accessibility *for* motor vehicles, but what matters is accessibility for travellers *by* motor vehicles.[16]

The consequence of this is that although *Traffic in Towns* convincingly demonstrated that the use of private cars would have to be limited, and correctly concluded that a vital question would be to determine what part should be played by different means of transport, its positive suggestions on this question are much less satisfactory. The clearest discussion of this point is not in *Traffic in Towns* itself but in an article written by Professor Buchanan shortly afterwards.[17]

We . . . always returned to the view that the motor vehicle was an incredibly useful method of transport, offering advantages possessed by no other invention to date, nor, as far as we could see, by anything in the foreseeable future. This led to our 'fundamental standpoint' of accepting the motor vehicle as beneficial and then seeking to understand what needed to be done to cities in a creative and constructive way to enable it to be exploited within them. The case for restriction would have to emerge, we argued, from the demonstrated impossibility, or great difficulty, or great expense, of meeting the full demand for the use of motor vehicles. I remain absolutely convinced this was a sound approach, and that had we started in any other way our study would have carried no confidence with the public.

how pedestrian movement between environmental areas is to be handled, nor about cyclists. However, in one of the practical studies an illustration is given of how pedestrian paths could cross the barriers represented by distributor roads. The position of cyclists is briefly discussed in the context of the same study.[15]

I would expect any local authority studying the problems of its area to proceed broadly as we did, that is, to investigate how far it is possible to go with the motor vehicle and to allow the case for restriction to emerge in the negative 'feedback' manner described. There is nothing to prevent the local authority at any stage costing out alternative methods of discharging the movements, provided they can think of them. In practice, this is likely to boil down to one main issue – how much of the *journey-to-work* load can be discharged by private cars, and how much must be carried by public transport.

The rule that this passage seems to suggest is that the use of motor cars should be maximised within the limits imposed by environmental, physical and financial constraints; given that there must be some restraint, then it must fall on optional rather than essential traffic. The role of public transport, which is referred to as if it were something quite distinct from motor vehicles, is to cater for the residual load of journeys which it is impossible to accommodate by car. This is a very negative view of public transport, which appears to be regarded as an inherently inferior system, that no-one who had an option would choose to use even for the daily journey to work. Even if this were correct, it is hard to see why the aim of maximising the use of cars should have priority over the aim of improving the basic transport system for people who do not own cars, and for those car owners who will not be able to use their cars. The smaller the proportion of journeys that it is possible to provide for by car, the more telling this objection.

The distinction which *Traffic in Towns* drew between essential and optional traffic also turns out to be much more complex than was first supposed. Originally the distinction seems to have turned on the importance of the journey; but then the idea of the relative ease of making the journey by different modes was introduced. It is presumably for this reason that work journeys, which are obviously of great importance, are classified in *Traffic in Towns* as optional traffic. In addition to these distinctions, some journeys have to be made to a particular place, whereas for others several destinations might do; some have to be made at a particular time, and others are more elastic in their timing. These distinctions were never properly sorted out, nor were they related to feasible forms of discipline; how can one be sure that the disciplines that might be introduced would discriminate in the way intended?

In passing, it should be noted that this same problem of finding controls that would discriminate exactly in the way intended also arises in connection with local and through traffic. In *Traffic in Towns* it is made clear that through traffic and longer distance traffic would have no choice but to use the distributor network. As a point of principle, this marks an advance, since it affirms society's right to control drivers' freedom to choose routes as they wish, if their unrestricted choice would conflict with other desirable objectives. But in practice it is not always easy to find ways of preventing through traffic or longer distance traffic from using local roads without at the same time making journeys for local traffic more awkward, and perhaps even increasing the volume of traffic on the streets it is desired to protect.*

Buchanan's interpretation of the term 'accessibility' is unfortunate in that it prevented him from developing the idea of a coordinated transport policy in a more satisfactory way. This same mistaken interpretation is also at the root of one of the most central ideas of *Traffic in Towns*, the idea that there is an inevitable, or at least a very general, tendency for the requirements of good environment and good accessibility to be in conflict. There is certainly a conflict between the environment and the presence of a large number of motor vehicles. But in any useful sense of the word 'accessibility' there may well also be a conflict between accessibility and the presence of a large number of motor vehicles. The larger the city, and the more that it is necessary and desirable to rely on public (or shared) means of transport, the more likely that such a conflict is. In these circumstances the disciplines required to protect the environment will coincide very closely with those required to safeguard the operation of a rational transport system. To rely on public transport is conducive to the environment in another way, too, since it is much easier to ensure that vehicles remain on the main road network rather

* The illustrative scheme for Norwich in *Traffic in Towns* shows how this might come about. The city is shown broken down into a number of 'rooms' by the imposition of barriers to vehicular movement at certain strategic points. These barriers would make it impossible for traffic to go right through the city, or right through any one of the 'rooms'. But traffic passing from one room to the next is obliged to go out to a distributor road on the periphery, round and in again. For some journeys, the mileage performed within each room under this arrangement would be very much greater than if there were no barriers.[18]

than cutting through streets where their presence is undesirable. Physical alterations are not required, at least on any scale. For buses, there is no problem at all, since their routes are specified. Even for taxis it would be possible to impose some discipline on the routes to be used simply by decree.

This should come as a great relief, since if the conflict between environment and accessibility were really as severe and universal as Buchanan suggested, and could really only be mitigated by spending very large sums of money, the prospects would indeed be bleak. They would be particularly bleak for those who care about the environment, since a drastic curtailment of accessibility is unlikely to come about and would be hard to justify, nor are the sums of money indicated by the practical studies of *Traffic in Towns* likely to be found, even if the environment would really be improved by the schemes suggested. Buchanan did not in fact recommend any particular level of expenditure; he thought it his function simply to illustrate what could be done assuming that different scales of physical reconstruction were possible, so that society could make a more informed choice. But even the most modest of the schemes illustrated involved considerable expenditure, and one of the conclusions of the whole study was that 'all the indications are that to deal adequately with traffic in towns will require works and expenditure on a scale not yet contemplated'.[19]

Buchanan's concept of a hierarchy of distributors has had a very important influence, even to the extent that a network which does not have a hierarchical structure is deemed faulty for that reason alone: to introduce a hierarchy becomes an end in itself. It has also been instrumental in encouraging the idea of urban motorways. The local roads, sub-arterial roads and arterial roads which Tripp had recommened form a hierarchy, but he had not taken the idea further than that, and even his arterial roads were not motorways: they were very often connected with sub-arterial roads by roundabout intersections, and they were supposed to carry buses and have bus stops, although the stops would be embayed.[20] The argument in *Traffic in Towns* for a hierarchy reads as follows:[21]

The need for a hierarchy of distributors. The function of the distributory network is to canalise the longer movements from locality to locality. The links of the network should therefore be designed for swift, efficient movement. This means that they cannot be used for giving direct access to buildings, nor even to minor roads serving the buildings, because the consequent frequency of the junctions would give rise to traffic dangers and disturb the efficiency of the road. It is therefore neccessary to introduce the idea of a 'hierarchy' of distributors, whereby important distributors feed down through distributors of lesser category to the minor roads which give access to the buildings . . .

This argument, however, only shows the need for two categories of distributor, exactly equivalent to Tripp's arterial and sub-arterial roads, and even then the argument depends on the idea that a town is best served by large volumes of fast-moving traffic.

The idea that some of the higher-category distributors should be motorways is argued as follows:[22]

Thus in many cases the links of a primary network as envisaged here would carry sufficient traffic to justify their being reserved for motor traffic only, with fly-over type intersections throughout. This is the specification which has come to be known in this country as a 'motorway'. We do ourselves, at a later stage in this report, refer to the need for certain distributors to be built to 'motorway standards' purely as a result of the volume of traffic they have to carry.

If there are to be urban motorways, then a good argument for a hierarchy of distributors can be made: that it is unsuitable for large volumes of motorway traffic, with drivers accustomed to travelling at motorway speeds, to be discharged straight on to local streets. But for this argument to apply, the motorways must first by justified quite independently of the idea of a hierarchy. The fact that to build motorways in towns requires the existing roads to be re-structured as a hierarchy is in itself an argument against motorways, not for them.

Traffic in Towns had great success in drawing both popular and official attention to the problems. It is no longer possible for environmental considerations to be disregarded as they had been. It also demonstrated the need to introduce controls over the way that roads are used, even though the positive suggestions made about the purpose and form of control are only partly satisfactory. There are many towns in which the ideas of *Traffic in Towns* would produce a satisfactory, if rather expensive, solution, as we shall see in Chapter 4, but they could only be relatively small towns. Large towns are not, as Buchanan's analysis seems to suggest, merely small towns enlarged. In large towns, the weakness in Buchanan's concept of accessibility becomes fatal, there is a real danger that the environmental losses in re-structuring the town in the way that his design principles indicate would exceed the environmental gains, and the costs involved even in modest solutions along the lines that he suggested are likely to be exorbitant.

Transportation Studies

The other major influence since the time that the traditional policy principles that have been discussed were formulated has been the transportation study. Studies of this kind originated in America in the 1950's and have now been conducted or are in progress in every conurbation in Britain and in many other towns. The aim is to predict future travel needs, demands or patterns (no very clear distinction is drawn between these concepts) and to select a plan for the town which will best serve the predicted demand. Such a study is a massive affair, which may take several years to complete, and there are important differences in the predictive methods of different studies. But in the present context, we are interested only in the new understanding of the problems and methods of attack that these studies have produced. For this purpose, a very broad and generalised account of the procedure is sufficient.*

The area under study is divided into zones, and the first step is to establish a complete picture of all the journeys made between the different zones in a given year. The main instrument for this purpose is a household survey conducted on the basis of a large sample. The journeys made in one day by each member of each selected household are recorded in an interview. Since the households interviewed have been selected so as to be representative of all the households in the area, and the interviews are properly spread over a year, the results of this survey can be grossed up to give the required picture of the journeys made in the survey year by all households in the area. Supplementary surveys are required to give the corresponding information on

* For a full account of the work involved at each stage of a transportation study, see *Introduction to Transportation Planning*.[23]

TABLE 2

Trip generation rates per household for an average weekday

HOUSEHOLD CHARACTERISTICS			JOURNEY PURPOSES	
Number of cars owned	*Income*	*Number of employed persons*	*Work*	*All others*
None	Low	0	0·02	0·58
		1	1·03	0·71
		2 or more	2·11	0·92
	Medium	0	0·02	1·17
		1	1·21	1·34
		2 or more	2·63	1·39
	High	0	0·03	1·43
		1	1·06	1·40
		2 or more	3·48	1·71
One	Low	0	0·02	2·10
		1	1·48	2·07
		2 or more	2·48	2·32
	Medium	0	0·01	2·91
		1	1·56	3·04
		2 or more	2·93	3·13
	High	0	0·02	3·02
		1	1·49	3·63
		2 or more	3·25	3·68
Two or more	Low	0	0·13	3·52
		1	1·21	3·79
		2 or more	2·71	3·82
	Medium	0	0·06	4·26
		1	1·38	4·91
		2 or more	3·34	5·01
	High	0	0·18	5·24
		1	1·66	5·83
		2 or more	3·43	6·02

Note: These figures and those of the following tables are illustrative only.

journeys made within the area by people who are not resident in households, for example, hotel visitors and commuters entering from outside, and similar surveys are undertaken to ascertain the number and pattern of journeys made by goods vehicles.

At the same time as this information on journeys is being collected, other information is sought which will help to explain the existing pattern of journeys and hence to predict the future pattern. For example, in the household survey various details about the characteristics of each household are collected, such as the number of employed people in the household, the total household income and the number of cars owned. Households are divided into categories on the basis of such characteristics and the average number of trips made per day is calculated for each category of household for each purpose of trip. Typical results from such an analysis are shown in Table 2.

To predict the number of journeys originating in each zone for the design year (usually about twenty years after the survey year), the assumption is made that trip-making rates established in the survey year for each category of household will hold good in the design year. What will change, however, is the number of households in each category. In particular, as people become richer and more cars are owned, there will be more households in the categories with high trip-making rates, and fewer in the categories with low trip-making rates. So the number of trips originating in any zone can be expected to increase, even if the number of households in the zone remains unchanged. This is illustrated in Table 3.

Given independent forecasts of the number of households in each zone in the design year, and the way that they will be divided between the appropriate categories, an estimate can thus be made of the number of trips that will originate in each zone in the design year. In similar ways, forecasts can be made of the number of trips that will terminate in each zone in the design year. For example, the number of journeys to work that will terminate in any zone clearly depends upon the number of work places that will be made available in that zone. The number of shopping journeys that will terminate in any zone is likely to be related to the shopping floor space to be provided there, and so on. The first stage of predicting future travel patterns for the design year takes the form of predicting the number of trips that

TABLE 3

Growth in number of trips generated by a particular zone for all purposes other than work

HOUSEHOLD CHARACTERISTICS			TRIPS PER DAY	SURVEY DATE		DESIGN DATE	
Number of cars owned	Income	Number of employed persons		Number of households	Number of trips	Number of households	Number of trips
	Low	0	0·58	179	104	55	32
		1	0·71	157	97	41	29
		2 or more	0·92	84	77	24	22
	Medium	0	1·17	51	60	22	26
None		1	1·34	67	90	30	40
		2 or more	1·39	82	114	38	53
	High	0	1·43	3	4	2	3
		1	1·40	9	13	5	7
		2 or more	1·71	28	48	13	22

One	Low	0	2·10	23	48	19	40
		1	2·07	32	66	25	52
		2 or more	2·32	25	58	16	37
	Medium	0	2·91	27	79	51	148
		1	3·04	36	109	88	268
		2 or more	3·13	47	147	101	316
	High	0	3·02	21	63	49	148
		1	3·63	33	120	65	236
		2 or more	3·68	56	206	116	427
Two or more	Low	0	3·52	1	4	2	7
		1	3·79	4	15	4	15
		2 or more	3·82	4	15	4	15
	Medium	0	4·26	4	17	12	51
		1	4·91	6	29	17	83
		2 or more	5·01	11	55	41	205
	High	0	5·24	5	26	26	136
		1	5·83	8	47	42	245
		2 or more	6·02	17	102	92	554
Total				1000	1813	1000	3217

will originate and terminate in each zone. This stage is usually
known as trip generation and the situation when that stage is
finished is shown diagramatically in Table 4.

The next problem is to predict how the total journeys from each
zone of origin will be divided among each possible zone of
destination – in other words to complete Table 4 by putting
numbers in each cell of the table in such a way that they add up
to the totals already established in the trip generation stage.

TABLE 4

The situation at the end of the trip generation stage of forecasting
non-work journeys

| Zone of origin | Zone of destination | | | | Total origins |
	A	B	C	D	
A					6411
B					6632
C					4412
D					3217
Total destinations	6370	6643	4349	3310	20672

Note: The number of zones in a real town is likely to run into scores
or hundreds.

This process is known as trip distribution, and the situation at the
end of that stage is shown in Table 5. Trip distribution is usually
accomplished by means of a mathematical formula which is
designed to reflect the way that different destinations compete
with each other. For example the employed population in any
zone will tend to divide itself among the different zones where
work is available in proportion to the number of jobs that each
zone provides. But workers will also take into account the difficulty
of travelling to each zone, which is usually measured by the time

required to travel there. The more distant zones, in terms of time, will attract the least number of trips. Trip distribution, therefore, takes account of those two considerations: the relative desirability of arriving at each possible destination, and the relative difficulty of travelling to each from a given starting point.

The time that will be taken to travel to each zone in the design year depends upon what roads it is intended to provide; estimates of travel time are made from the road plans that are being tested.

TABLE 5

The situation at the end of the trip distribution stage of forecasting non-work journeys

| Zone of origin | Zone of destination | | | | Total origins |
	A	B	C	D	
A	—	3326	1184	1901	6411
B	3482	—	2533	617	6632
C	1076	2544	—	792	4412
D	1812	773	632	—	3217
Total destinations	6370	6643	4349	3310	20672

But the precise formula to be used in distributing trips in the design year is established by seeing what produces the best fit to the travel pattern observed in the survey year.

When the forecast of the geographical pattern of journeys has been completed by trip distribution, the next stage is to predict what modes the different travellers will take. In other words the single origin and destination table for all personal journeys is broken down into several tables, one for each mode, as is shown in Table 6. Again this is done by trying to establish a formula from the survey to explain how travel is divided between the competing modes in terms of the performance of each (time, cost,

TABLE 6

Modal split of non-work journeys

1. Journeys made by car drivers

Zone of origin	Zone of destination				Total origins
	A	*B*	*C*	*D*	
A	—	941	357	630	1928
B	989	—	1479	357	2825
C	324	1493	—	520	2337
D	585	471	409	—	1465
Total destinations	1898	2905	2245	1507	8555

2. Journeys by car passengers

Zone of origin	Zone of destination				Total origins
	A	*B*	*C*	*D*	
A	—	380	129	264	773
B	403	—	617	149	1169
C	129	602	—	209	940
D	253	198	172	—	623
Total destinations	785	1180	918	622	3505

3. Journeys by bus

| *Zone of origin* | *Zone of destination* | | | | *Total origins* |
	A	*B*	*C*	*D*	
A	—	2005	698	1007	3710
B	2090	—	437	111	2638
C	623	449	—	63	1135
D	974	104	51	—	1129
Total destinations	3687	2558	1186	1181	8612

Note: It would be common practice to add together the journeys made for all different purposes before dividing them up between competing modes, but this step is left out for the purpose of the illustration.

frequency, etc). The same formula is used for the design year, but what the performance of each mode in the future will be is estimated from the proposals under consideration.

As a result of the modal split stage, what was previously an origin and destination table showing personal journeys can be expressed as an origin and destination table showing vehicular journeys. The final stage of prediction is to forecast what routes road vehicles will take through the road network and hence to calculate the volume of traffic on each road in the network. (Special problems arise when one of the competing modes is rail or underground, but they need not concern us.) For this purpose, some simple rule is adopted such that each driver will follow the route which is shortest by distance, time or some other criterion. But the procedure adopted usually allows that this route will not hold absolutely – some traffic is allotted to routes other than the shortest. This final predictive stage is known as traffic assignment; a simple illustration of it is given in Figure 1.

The assignment procedure also constitutes the test of the road

Figure 1   Traffic Assignment

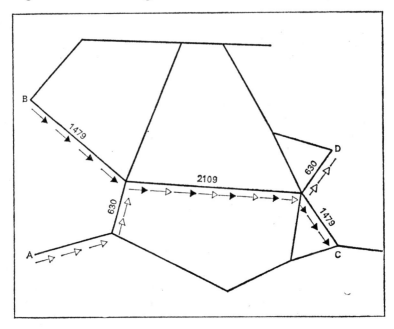

network under consideration. The flows assigned to each road can
be compared with its theoretical capacity. For the network to be
judged satisfactory, it is necessary that the theoretical capacity is
not exceeded, or is exceeded by a small amount on only a few
roads. But if on certain roads the assigned flows were very much
less than the theoretical capacity, those roads would be regarded
as unnecessary, or at least unnecessarily wide.

Transportation studies hardly ever contain an explicit analysis
of the problems of the town under study. But there is a very
particular view inherent in the whole method of approach that
has just been described, even though very few practitioners seem
to have realised quite what they have implicitly committed
themselves to.

The essence of the study is that travel made in a particular
town at a particular time is surveyed and is related to the size and
characteristics of the population and, to a lesser degree, to the
existing transport facilities. The relationships found are then
applied to the future population in some design year to forecast

future travel. Deeply embedded in this whole procedure is the idea that the present situation is satisfactory. Why project indefinitely into the future relationships observed in a thoroughly unsatisfactory situation?

On the other hand why do a study at all if the present situation is desirable – there is nothing to put right? The answer to this is presumably that although the existing facilities may be adequate for the present situation, the situation is changing and they will not be adequate for the future. However, the present situation is implicitly set up as a target for the future, in the sense that planned facilities are judged satisfactory if they give the same service to the future population as the existing facilities give to the present population. The study may indicate that this target cannot always be achieved, but it still remains the relevant ideal; failure to reach the ideal is always regrettable and any significant departure from it would render the plan unacceptable.

It is ironic that transportation studies should idealise the existing situation in this way, since if there had not been considerable dissatisfaction with it the studies would never have been undertaken. The explanation lies in the American origin of the study methods. The chief problem, in the early American studies at least, was urban growth. The Chicago study is usually considered to be the first major transportation study. In Chicago, the population was expected to increase by 50% between the survey year and the design year, and the area of land in use by nearly 100%. It was obvious that the existing transport system would be inadequate, simply in terms of its geographical extent. Rightly or wrongly, it was not thought necessary to change the type or character of facilities – the problem was simply to provide more of the same, so that the traveller of the design year could enjoy the same level of service as the traveller of the survey year.

The way in which the existing situation figures as a target is most obvious in the trip generation stage. The number of trips made is taken to be determined only by the number and characteristics of the households in the area, without any reference to the facilities provided, whether roads or public transport. According to the formulae used, no improvement in facilities could generate a single trip and no worsening of facilities could suppress a single trip. Of course, no-one would suppose, particularly if the population were increasing, that the same number of trips would be made whether any new facilities were provided or not. But the procedure is intelligible if the plans under test are *designed* to provide future travellers with the same level of service as that now enjoyed by present travellers. In the trip generation stage, it is implicitly assumed that the plans will succeed in their aim; only when all the stages are complete, and the flows assigned to each road in the network are compared with its capacity, can it be seen whether they really will succeed or not. But the aim is itself a sensible one only if the present situation is satisfactory. It would be the height of folly to plan to perpetuate a situation that no-one approved of.

The assumption that the present situation is satisfactory enough to act as a target appears in a less strong form in the other stages of the study, to a degree which depends upon the particular methods used. It also explains some of the surprising omissions of the studies. No attempt is made to find out what improvements travellers would like to see in the conditions in which their present journeys are made, although there is plenty of independent evidence that travellers are dissatisfied. Neither is any consideration given to the present frustrated demand – to the problems of those who would like to make journeys but are deterred by present conditions from doing so. But of course, if the present situation really were satisfactory, there could be no significant dissatisfaction with travelling conditions and no significant frustrated demand.

Transportation studies completely ignore the interests of pedestrians and cyclists. They do not even attempt to collect a full record of the journeys made on foot or by cycle. This neglect is again presumably accounted for by the American origin of the studies, and it accords with the tendency to disregard these travellers that has developed under the pressure to cope with increasing volumes of motor vehicles. But this tendency is itself regrettable and it is most unfortunate that a bias of this sort which has grown up in practice should now become formalised in the very methods of study used.

There is nothing in the predictive procedure as outlined which itself tends to favour cars as against public transport, but when the procedure is applied to current British problems it produces a marked bias against public transport. Public transport users have more reason to be dissatisfied with the existing travelling conditions than have private motorists, and the studies themselves have produced evidence which suggests that there is more frustrated travel demand among those who are entirely dependent on public transport than among motorists. So to ignore these problems has a more damaging effect on the case for providing public transport than on the case for providing facilities for motorists. Also, when the modal split models are examined in detail, they are usually found to contain some bias against public transport. In particular, because predictions are made throughout in terms of households rather than individual people, there tends to be an assumption that everyone in a car owning household has a car available for each journey that he wants to make. This ignores the problem of the large class of people in car owning households who do not have the use of the car, because they are too young, too old or otherwise unable to drive a car, or because someone else in the household is already using it.

But the most important bias of all, in the British context, is contained in the central assumption that has already been discussed, that the aim should be to provide tomorrow's traveller with the same level of service as today's. This aim may be quite difficult to achieve for public transport passengers if it is assumed, as transportation studies do, that the market for public transport will naturally contract so that there will be many fewer passengers to support the service. That particular difficulty, however, is usually assumed away in the predictive techniques adopted, which presuppose that fares and frequencies will not be affected by the decline in custom. The major difficulty comes in providing the same level of service for tomorrow's motorist as for today's, since there will be so many more car owners than there are today and facilities for cars are so very expensive. By adopting this aim, one is in fact committing oneself to open-ended and possibly enormous expenditure on one particular form of transport before any case has been made out for it. A criticism made of the London study applies to most other studies too, that it 'postulated the general nature of future investment in advance of the study of future investment needs'.[24]

As the outline account given has shown, transportation studies are a way of testing out different plans to see whether they are adequate, particularly in terms of road capacity, to satisfy certain implicit objectives.

The studies contain no method of formulating plans, which must be drawn up previously and independently. How they are formulated is obviously crucial, since even the best method of testing can only select the best of the plans submitted to it. Clear accounts of why particular plans were chosen for testing are often lacking, but they must be formulated in the light of the planner's analysis of what the important problems are, even if his analysis is not very coherent or explicit. Given the historical background which has been sketched, and the way that the transportation study method itself tends to focus attention on one particular aspect of the problem, it is not surprising that new road construction is usually the dominant feature of plans tested, and that rules for the use of the roads and other transport facilities have not been tested.

To summarise, it may be said that in terms of thought about policy, transportation studies have provided no new principle. The underlying principle is still that all demands should be met, or that the anticipated growth in traffic should be provided for. This principle is not made explicit, but is deeply embedded in the whole approach and method adopted.

Transportation studies have also given a new lease of life to the mistaken use of the term 'demand'. At least the first three predictive stages, trip generation, distribution and modal split, are usually referred to as forecasts of demand. If, when the assignment is complete, it is found that there is some overloading on the proposed network, this is taken as a measure of the extent to which future demand would be restrained if that plan were adopted. There is no justification for this use of the term.

Transportation studies start with the use made of the existing facilities and attempt to forecast what use would be made of certain proposed facilities. At no point do they endeavour to discover what the demand is, if by this is meant how people would like to behave as opposed to how they actually behave. This is not a trivial matter of the mistaken use of technical terms; it has a profound bearing on policy. The implication is that unless the demands are satisfied, or nearly so, the planner will have failed to meet the legitimate aspirations of the people for whom he is planning. This raises once more all the points discussed in Chapter 2. What people's aspirations are cannot be ascertained simply by recording and projecting facts about their present behaviour, even if the record were complete and the projection unbiased. To identify what people would like to do with what they actually do is to assume, once again, that the situation in which their behaviour is observed offers them a satisfactory range of choice. In other words, it is to assume away all the problems about which there has been so much public concern and which are the stimulus of the studies.

Where transportation studies have produced great advances is on a technical level. We can now survey, and represent on computers, complex land use patterns, transport networks and movements and to a limited extent we can predict the future. Most important of all, we have the ability to test the capacity of transport facilities to deal with certain hypothetical loads. To see how much of an advance this is, one has only to consider the situation of the traffic engineer before these techniques were developed. His job was to predict the growth of traffic and to provide for it. But how was traffic to be predicted, and how was the engineer to tell whether or not any proposed scale of provision was adequate? The favoured method of prediction was apparently to count traffic on particular roads and to apply some kind of growth factor, related to the expected increase in the number of vehicles.[25] It was known that this method was not satisfactory, since it failed to take account of how movement was related to the pattern of land use, or how traffic spread itself over a whole network of roads, or how private and public transport competed with each other. But the engineer had no formal means of taking these effects into account, he could only rely on his judgement. Similarly, although he could tell whether an individual road or junction could handle a predicted traffic load, he could not tell whether a whole network could handle a complex pattern of flows. Hence he had no means of testing the capacity of a larger plan, such as a town's road system.

The predictive power of transportation studies should not be overrated; they do not take full acount of traffic generation, and even apart from that, the difficulties of predicting in the long term such a complex and subtle phenomenon as the complete pattern of movement in a town are obviously formidable. It would be unwise to expect the results ever to be anything better than highly approximate. Their great contribution is that they enable us to explore the potential of different policies. These policies need not be merely different road plans, they could also include management measures.

For example, transportation study methods could be used to provide at least rough answers to questions such as these: 'Suppose we could find some way of diverting all the through traffic now going through this residential area, do the perimeter roads have the necessary capacity to handle it?' 'Suppose that all car-borne commuters were obliged to change to public transport at certain specified interchange points at the outskirts of a city, what would the effect be on the time taken by all travellers?'

So we can borrow techniques from the studies, and adapt them to our own purposes, even though the complete studies as now conducted contain a systematically distorted appreciation of the problem.

Other developments

Both *Traffic in Towns* and transportation studies have helped to provide evidence of the impossibility of continuing to follow policies based on the precept that 'all demands should be met'. Neither produced very satisfactory alternatives, either as to how roads should be used, or as to how new road building should be justified. But there have meanwhile been other attempts to consider these questions which should be briefly mentioned at this point and will be discussed again in later chapters. Economists have long been unhappy about the criteria used to justify investment in roads, which differ so markedly from the criteria used in other fields, and have been trying with some success to formulate criteria based on something more than the simple test of capacity. In other words they have been trying to find satisfactory ways of answering the question: 'how much of our resources should be devoted to building roads, or providing other transport facilities, given the other claims for the use of those resources?' They have also been concerned about the inefficiency in the way in which roads are used, and have proposed a remedy based on the idea of the user paying for the roads while he uses them, the amount to be paid being related to the degree of congestion prevailing. Road pricing is a valuable idea, but even more valuable has been the fact that by drawing attention to the waste and irrationality inherent in our present indiscriminate use of the roads, economists have helped to ventilate the whole question of possible alternatives.

The other important development in very recent years has been a series of officially encouraged experiments in managing the use of roads for purposes other than the traditional purpose of accommodating more motor vehicles. Some schemes have had environmental aims, to protect residential areas for example, others have been concerned to help the pedestrian and others to facilitate the use of the bus. In one way or another they have all challenged the traditional implicit priorities that travel is more important than other activities and the motorist more important than other travellers. Some of these ideas are described in Chapter 9.

4. Oxford

Of all British town planning disputes, that of Oxford is the most famous, protracted and emotionally charged. It has involved writs in the High Court, debates in the House of Lords, discussions in the Cabinet and all kinds of wrangles between town and gown and between different groups of colleges. An Archbishop of Canterbury has maintained that 'the real worth and integrity of this nation will be judged in future ages by what we do or do not do to rescue Oxford'.[1] Fortunately, many of the complexities can be ignored for the purposes of the present account. The political and legal arguments are of no interest, and even the planning history can be treated in quite a summary way, since the aim is not to give a narrative but to set out and appraise the main arguments that have been used, particularly those that have weighed with the people responsible for taking decisions. The easiest way to do this is to tell the story in rough chronological order, but the reader should be warned that many events have been left out, and that all the suggestions that are discussed here, and a host of others besides, had been made long before the time that they are first mentioned in the present account.*

The problem that has been causing concern at Oxford since before the Second World War is that of traffic in the university area, and the High Street in particular, which is quite inconsistent with the conditions that ought to obtain in a university or in one of Europe's finest streets. The need to reduce the traffic drastically has not been disputed; the argument has centred on

---

* Anyone interested in pursuing the story in more detail should start with *Oxford Replanned* by Thomas Sharp, published for the Oxford City Council by the Architectural Press in 1948. The *Report of the Oxford Road Enquiry* published by the Ministry of Housing and Local Government in 1961, besides being an important document in itself, contains a useful summary of events up to that time.

how best to do it. For a long time the issue was obscured by the belief that the problem was one of traffic which had no business either in the university area or elsewhere in the city, but was going through Oxford only because Oxford happens to lie

Figure 2   Oxford main features

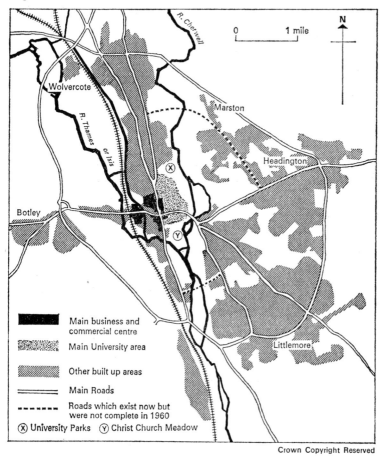

Crown Copyright Reserved

athwart a number of traditional main roads. It was, therefore, believed that the provision of a complete set of ring roads would be a sufficient remedy. This belief was incorrect, as many people had always realised, but it was not finally demonstrated to be incorrect until an origin and destination survey was undertaken in 1957.[2]

As this survey showed, most traffic in Oxford is generated by Oxford. The most important source of attraction is the commercial centre, which lies to the west of the university. Traffic between the commercial centre or adjacent areas and east Oxford comes

Figure 5   Place names in central Oxford

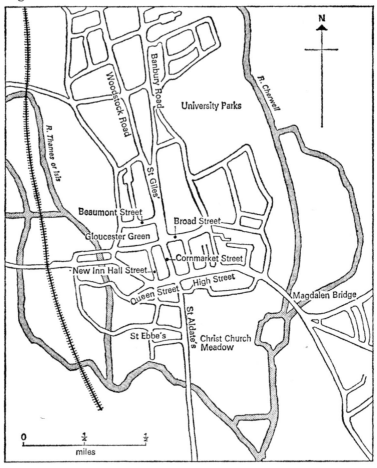

along the High Street because Magdalen Bridge is the only convenient river crossing.

Given, therefore, that by-passes would not suffice, where was a new road to relieve the High Street to go? Roads both to the north and the south were suggested, but both suggestions aroused considerable protests, principally because of the damage that would be done to the University Parks and science area on the north or to the fine open space of Christ Church Meadow on the south, but also because of other damage elsewhere on the routes. After a long period of dispute, a traffic survey – the one just referred to – was undertaken. This formed the basis of an analysis of the traffic consequences of a number of alternative road schemes which was carried out by the Road Research Laboratory. A special Inquiry, rather outside the normal planning procedure, was held at Oxford in December 1960, with the object of making an impartial and objective assessment of all the contending proposals.

Twenty-two plans in all were put forward to this Inquiry, but they can be considered as variations of three basic designs. Each of the three involved a new road on the southern side; the distinction between them was in terms of the precise location of the route. The most southerly route crossed the rivers just below their junction and went from there across to the railway before turning north. It thus avoided crossing either Christ Church Meadow or any part of the old city. After running along the railway for about a mile and a half, the road turned east along a route which again lay outside the city centre or the university area and continued across the Cherwell back to East Oxford. This whole plan was known as Scheme A. Scheme B involved a road as close to the High Street as possible with a new bridge some five hundred feet south of Magdalen Bridge, a road across Christ Church Meadow, over St Aldate's, where a new roundabout was proposed, turning north through St Ebbe's, over Queen Street and finally up to meet St Giles' at the corner of Beaumont Street. From the St Aldate's roundabout, a new road running west was proposed finally merging with the existing roads which gave access to the station. Under this scheme, Magdalen Bridge would have been closed to motor vehicles. The scheme proposed by the City Council was intermediate between these two. It involved a route through the middle of Christ Church Meadow and across to the railway where it continued along the line proposed by Scheme A as far as the Woodstock Road. A branch of this road ran up through St Ebbe's parallel to St Aldate's to a junction with Queen Street.

Both Scheme B and the City's scheme were opposed on the grounds that any road across Christ Church Meadow would

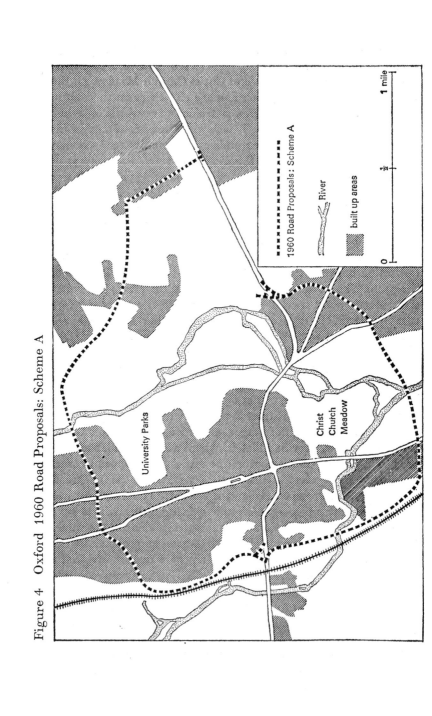

Figure 4   Oxford 1960 Road Proposals: Scheme A

University Parks

Christ
Church
Meadow

1960 Road Proposals: Scheme A

River

built up areas

0   ½   1 mile

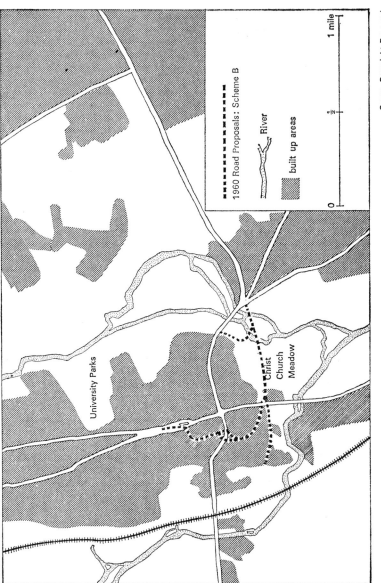

University Parks

Christ Church Meadow

1960 Road Proposals: Scheme B

River

built up areas

0   ½   1 mile

Figure 6   Oxford 1960 Road Proposals: City Council's Scheme

1960 Road Proposals: City Council's Scheme

River

built up areas

0   ½   1 mile

University Parks

Christ
Church
Meadow

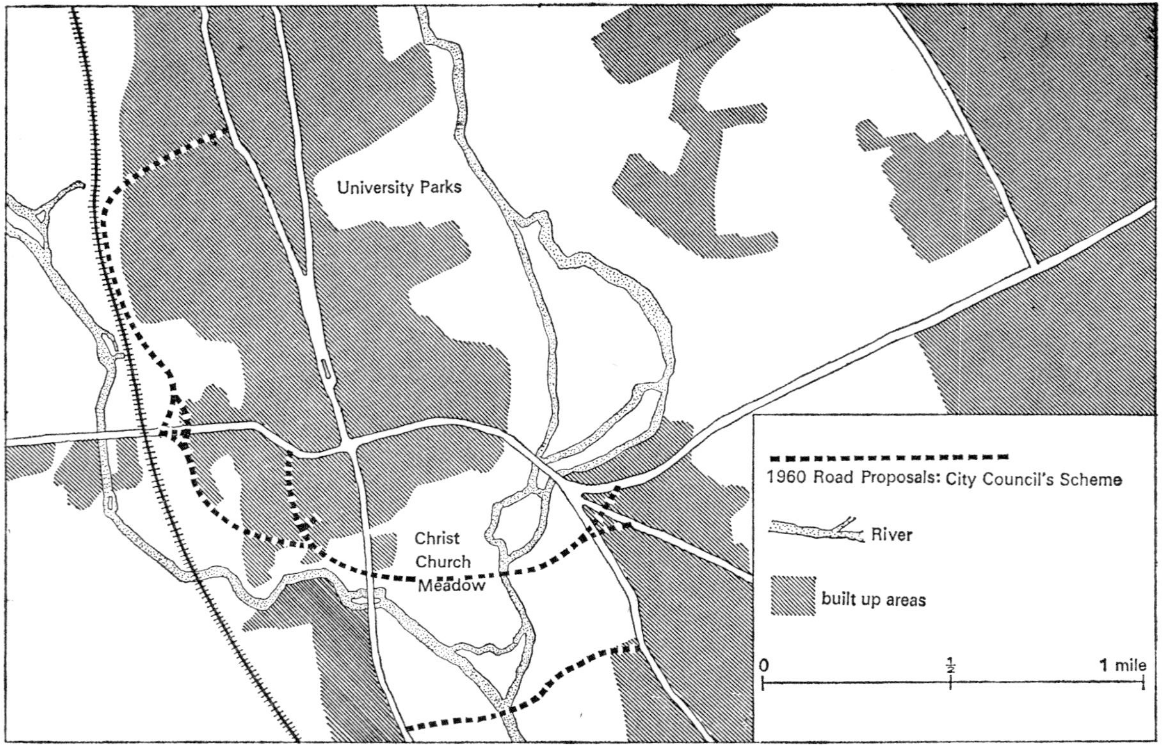

destroy its character, but this was by no means the only objection. Both involved damage to open space and property on the eastern side of the river. Both involved major roads through St Ebbe's. Although much of St Ebbe's was due for redevelopment anyway, so that the demolition of existing properties was not as serious a consideration as it otherwise would have been, new roads through this central area would have been hard to incorporate in any new development. The link in Scheme B between Queen Street and St Giles' was also a very damaging feature which involved much demolition of property. In fact the physical damage done by Scheme B and its variants in various parts of the city, even disregarding the Meadow crossing, were such that they attracted virtually no support at the Inquiry, even though the Road Research Laboratory's assignments showed that they would have been much the most effective in taking traffic out of the university area.[3] In addition, the City Council took the view that a physical restriction such as the closure of Magdalen Bridge was undesirable in principle.[4]

The City did not deny that their own scheme would do damage, especially to Christ Church Meadow. Without agreeing with those objectors who said that the Meadow would be a complete loss, the City nevertheless said that they would gladly have left it alone if they could have found an effective alternative route. But Scheme A did not provide such a route. It was too far out to attract the motorist, hence it would not provide sufficient relief even in the short term; and in the long term, with the natural increase of traffic, the central roads would fill up again.[5] The City's counsel expressed the point as follows:

We should suggest that it is a very unattractive road for motorists and what one has to consider in any road of this sort, unfortunately, is that if you want motorists to use a road, since they cannot be dragooned into doing what they do not want to do, you have got to encourage them to do so by making it more convenient than what they have been doing before.[6]

The City also pointed out that even if Scheme A did no damage in the city centre or the university area, it did a great deal elsewhere, both to residential areas and to public open space.[7]

The Inspector accepted the City's case. He concluded:

It is with regret that I have come to the conclusion, on weighing the evidence, that Scheme A would not provide a practical solution of

c

Oxford's problem, and that, if peace and quiet is to be restored to the central university area, a road across Christ Church Meadow is inescapable. The loss of the long preserved special kind of amenity provided by the Meadow, and the provision instead of the lesser amenity that would no doubt be secured, will be regarded by very many people as a grievous loss, but against that loss must be weighed the great gain of peace and quiet in the ancient city that in my judgement cannot otherwise be obtained.[8]

As a further argument against Scheme A, the Inspector said that what benefits it would bring would 'be secured at too heavy a price in the destruction of houses, interference with the amenities of residential areas, and inconvenience, hardship and damage to a very large number of people'.[9]

The Minister agreed in the main with his Inspector, and the City went on to prepare its Development Plan broadly on the basis of the scheme that they had advocated at the Inquiry.* There the matter might perhaps have rested, if the following paragraph had not appeared in *Traffic in Towns*:

As to 'relief', it seems to us that relief roads are often designated without safeguards to ensure that the general increase of traffic does not soon make conditions as bad as ever on the relieved road. This is a special danger when the relieved road has some innate attraction for traffic. This point is so important that we are tempted to quote the controversial case of the High Street at Oxford as an example. In this case a relief road has been planned on an alignment which it is hoped will be sufficiently 'attractive to traffic' to give substantial relief to the High Street. The risk is that traffic will continue to use the High Street, and that only a measure of congestion in that street will force some traffic onto the relief road. As we have previously suggested, our approach would be to assess the 'environmental capacity' of the High Street, and then to consider what steps would be needed to reduce traffic to that level, and to hold it there permanently. Such steps would almost certainly include the compulsory direction of traffic, or most of it, onto the relief road. There would then be no

* At the Minister's insistence, however, the road which the City had proposed should stop at Queen Street was carried on via New Inn Hall Street and Beaumont Street to St Giles'. The reason for this was that it appeared from the Inquiry that unless some such extension was made there would be extreme traffic difficulties at the junction of the new road with Queen Street. The irony was, however, that this amendment made the proposal road almost as damaging as the rejected Scheme B.

particular need to choose an alignment for the relief road which would be competitive in journey-time with the old road, it could be put anywhere suitable. In the conditions that are going to arise in the future, as vehicles multiply in numbers, we think this kind of strict discipline of vehicular movement is inevitable.[10]

In other words, as a means of taking traffic from the High Street the Meadow Road was both unnecessary and insufficient. Unnecessary because once it has been admitted in principle that it is legitimate to control the routes that traffic may take, such a close alternative to the High Street is no longer required; insufficient, because unless controls are employed even a Meadow Road would not be attractive enough to be sure of diverting most of the traffic from the High Street.

It is the second point, that not even a Meadow Road would be sufficient to relieve the High Street, which is the crushing argument against the objection to closures on principle that the City expressed at the 1960 Inquiry. The High Street is a wide convenient street and the route between east Oxford and the city centre of which it forms part is nearly straight. It is obvious from a glance at the map that it is not possible to find an alternative route which would be clearly and substantially more attractive. Unfortunately, this evident fact had been obscured by the method of assignment used at the 1960 Inquiry. This was the first time that an assignment of this kind had been done in Britain, and the method used was quite unsuited to make predictions of the fineness required. In fact, the method was of the same type as that illustrated in Figure 1, page 50. Accordingly, if one route between two places was shown to be shorter than another by only a few seconds, all the traffic concerned was allocated to it, instead of being split between the competing routes more or less evenly, as one would expect in real life. The effect was that a quite misleading picture of the likely efficacy of the Meadow Road was created.

The next Inquiry, held under the normal Development Plan procedure, took place at Oxford in January 1965. The points just raised and other similar points were argued by Professor Buchanan appearing as a witness on behalf of the University. Buchanan and other witnesses also stressed the physical damage that the City's plans would bring about in other parts of Oxford.

The Inspector was sufficiently impressed by the evidence to

recommend that the entire plan be remitted to the City Council for further consideration. The Minister (by then Mr Crossman) overruled this recommendation of the Inspector on the grounds that much of the development proposed by the Plan was not dependent on the Road proposals and should be allowed to go ahead. But he recommended that consultants should be appointed to consider alternative routes within the context of a comprehensive plan for a road and traffic system in central Oxford.

The study which the Minister had asked for was duly commissioned. It took the form of a transportation study of the general type described in Chapter 3, but with two unusual and desirable features. An environmental study of the centre was carried out with the object of defining the maximum vehicle flow and amount of parking that should be permitted on each central street to conform with various environmental criteria, such as pedestrian convenience and the limitation of intrusion by vehicles on the visual scene. Also, early on in the study an estimate was made of the funds that could be made available for the purpose of investment in transport in Oxford, so as to avoid formulating and testing schemes which would not be financially feasible.

The study represented a great advance in another way too, that the inescapable need for some control over drivers' choice of route, and hence for closure of some central streets, was accepted from the outset. A common feature of all the designs tested was that Cornmarket and Queen Street would become pedestrian precincts, through which, however, some bus travel might be permitted.

The consultants examined two schemes which entailed no new road building other than some minor improvements already agreed to. The second of these schemes involved closing Magdalen Bridge to private cars. All the other designs tested involved new roads and river crossings to the south of Oxford. In one of the designs the road ran through the middle of Christ Church Meadow on the line of the road proposed by the City in the 1960 Inquiry. In other designs, the road took the line to the south of Christ Church Meadow that had been proposed as part of Scheme A in 1960.

A design with a river crossing much further south was also tested. But in contrast with the 1960 proposals, none of the designs involved new roads entering central Oxford from the

Figure 7   Oxford Recommended Road Plan

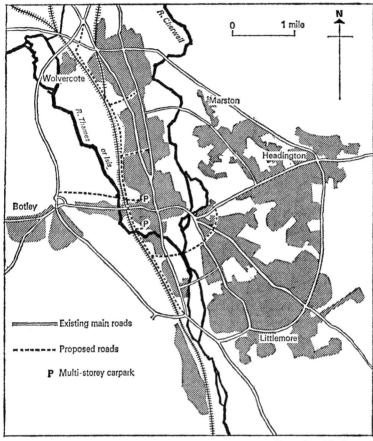

Notes
The roads on the central part of this map are the same as those for the approved plan, shown in Figure 8.

south. It had always been admitted that such roads would be damaging, but they had previously been proposed as part of the vain attempt to create a relief road more attractive than the High Street. But with the High Street in effect closed to traffic bound for the centre, there was no further need to include them. Instead, in each of the designs the route across the river was continued to join a new road running along the railway, and it was envisaged

that the main access to the commercial centre would be on
the west of the city from this new road. Spur roads were to
be provided from it to two new multi-storey car parks one in
St Ebbe's and the other by Gloucester Green. The design recommended by the consultants and adopted (with minor modifications)
by the City was one of those which in its river crossing corresponded
with the previous Scheme A. This design is illustrated in Figure 7.
After an Inquiry held in Oxford in 1970, the Inspector nevertheless recommended in favour of the road through Christ
Church Meadow. He believed that even with the closure of
central streets suggested a road further out might not be effective
in relieving the centre, and he was also impressed by the damage
a road on the proposed line would do, particularly to public open
space in south Oxford.[11] The Secretary of State, however, (by
now Mr Walker) ruled in favour of the City rather than his
Inspector on this point, although accepting other modifications
put forward by the Inspector. The design finally approved by the
Secretary of State is illustrated diagrammatically in Figure 8.

Although the design now approved is an improvement in
many ways over what the City proposed in 1960 or 1965, there
remain serious doubts whether it will be an altogether satisfactory or attractive plan. In the first place, it is not clear that the
primary aim of reducing traffic in the university area and the
other fine adjacent streets will be achieved.

Certainly there will be a very great reduction in the High
Street itself. But a number of streets on which traffic should be
held to very low levels are still part of possible traffic routes. On
some of them, the consultants' predictions suggest that the
desirable maximum flows defined in their environmental study
will be exceeded,[12] and there are several reasons for supposing
that their predictions may have underestimated the likely flows
on such streets. The consultants pointed out that the fulfilment of
the predictions depends on a reduction in the number of privately
controlled parking spaces in central Oxford and it is not clear
whether this reduction can be achieved.[13] Even if it can, the
method of traffic assignment used by the consultants was still one
which assigned all the traffic between a given origin and destination to the shortest route (shortest in terms of a weighed function
of time and distance) even though this route might be only
fractionally better than the second shortest. Detailed figures are

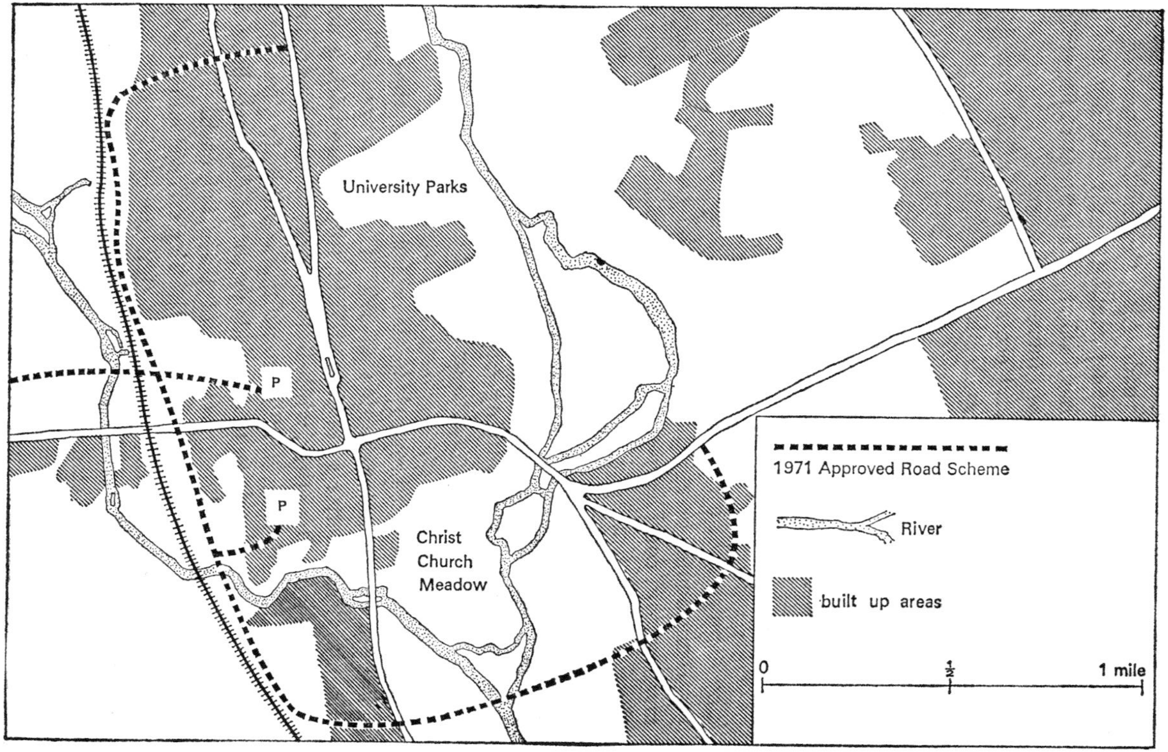

University Parks

Christ Church Meadow

P

P

1971 Approved Road Scheme

River

built up areas

0   ½   1 mile

not given in the report, but it seems reasonable to suppose that this unrealistic assumption would overestimate the diverting effect of the proposed relief roads. Also, the methods used in the predictive stages of the transportation study leading up to assignment are unlikely to have taken full account of the traffic generating effects of providing more road space.

The proposed car parks and access roads in St Ebbe's and at Gloucester Green attracted much adverse criticism at the Inquiry, particularly on aesthetic grounds. One of the modifications stipulated by the Secretary of State was that the Gloucester Green car park should be reduced in size: even so, both will be unsympathetic structures and will prejudice the development of parts of Oxford which, although now neglected, have great potential.

Finally, there is the effect of the southern relief road itself. The damage that a road on the proposed alignment will do has been commented on again and again over the years, and has been an important consideration in leading many discerning people to the view that a road across Christ Church Meadow would be the lesser evil. Although lovers of the Meadow, of whom I am one, will not agree with that judgement, one thing that has become absolutely clear in the long history of this dispute is that any new major road linking east Oxford with the commercial centre, whether it goes north or south, must be very damaging or expensive or both.

Could the present proposals be described as making the best of a bad job, or is a better solution possible? The only way to achieve a substantially better solution would be to dispense with a new east–west road altogether, while finding some other way to reduce traffic in the High Street. The consultants paid little attention to this possibility because to have done so would have been in conflict with two basic premises of their approach. The first was that the land use proposals of the city's Development Plan had to be accepted; the second was that traffic of roughly the volumes predicted would have to be accommodated, particularly in the commercial centre, if the Plan was to work. But both premises are very much open to question.

The way that the study was set up by Mr Crossman virtually precluded an examination of alternative land use plans. This was a pity, since an unfortunate disposition of land uses is the main underlying reason for Oxford's traffic problems. The university area and the beautiful open spaces along the Cherwell and the lower Thames lie between the residential districts of east Oxford and the city's commercial centre. This creates a pattern of journeys which is inconvenient for the travellers themselves as well as being damaging to the environment. It would be desirable over the years to build up centres in east Oxford (whether the existing centres or new ones) so as to reduce the need to travel between the two parts of the town. However the land use proposals envisage that the present centre will grow substantially and will continue to hold its predominant position.[14]

An alternative policy need not entail the decline of the present centre in absolute terms. It could remain the main centre for all that part of the town west of Magdalen Bridge and for outlying areas to the north, south and west. A new road along the line of the railway would probably still be required (whether or not it should be continued as far as the by-passes is an open question), but not a new east–west road.

This does not mean that the residents of east Oxford would be cut off from the existing commercial centre or from the university area. Direct access by car along the line of the High Street would have to be prevented in order to reduce traffic flows as much as possible, but direct access on foot, by bicycle (very important in Oxford) and by public transport would still be possible and would indeed be more attractive than at present. Those who wanted to go by car would still be able to, by taking a long way round, but the time advantage in favour of public transport would be so great that only those who had a compelling reason to use a car would do so.

The second premise can be questioned, even if the proposed land use plan is accepted, on the grounds that the movement of people that would be required to make the Plan work need not entail such heavy reliance on the use of the private car; the main alternative would be to use buses much more. The use of cars could be restricted by street closures, of the type envisaged in the present Plan but more extensive in scope, or by parking controls, or both. The bus system would not necessarily be limited to the present type of service, but might include new services such as a shuttle service across the city centre. This kind of solution has had its advocates at Oxford for a long time. A scheme of this type, worked out in some detail, was presented by an objector at the 1965 Inquiry and withstood considerable scrutiny by way of cross-examination.[15]

Many objectors at the 1970 Inquiry advocated a similar approach. However, it was rejected by the consultants, the city and the Inspector for very interesting reasons.

The consultants' study did not include an investigation of how a new bus system of this sort might work. What the consultants did do, however, in the course of their consideration of solutions involving no new road building, was to calculate what percentage of future journeys would have to be made by public transport in order for any such solution to produce flows consistent with the environmental criteria established for certain central streets. They concluded that for the study area as a whole, 38% of journeys (excluding those made on foot or by bicycle) would have to be made by bus, and that for trips to the centre the proportion would be 64%.[16] They maintained that this degree of dependence on public transport would be socially unacceptable and hence that to adopt a scheme that presupposed it would be fatal to the well-being of the business centre.[17]

These percentages would presumably be somewhat reduced if some new road building, such as the road along the railway, were included in the design. But the most interesting thing is how the consultants should have arrived at the crucial judgement that the estimated degree of reliance of public transport would be unacceptable. It was not based on any kind of survey of public opinion in Oxford or elsewhere, nor was any evidence from the experience of other city centres adduced in its support. The dependence on public transport envisaged would really only have altered the present pattern of journeys to the centre; elsewhere in Oxford the car would have retained its overwhelming position *vis-à-vis* the bus. The evidence produced at the 1970 Inquiry (which of course was not available to the consultants at the time that their report was being prepared) shows that there is a substantial body of opinion in Oxford actively in favour of restrictions on the entry of cars to the centre.

All of this is still consistent with the idea that there would be some risk to the prosperity of the city centre, at least to its shopping activities, if a high level of car use were not permitted; and a cautious policy maker might decide that even in the absence of positive evidence he was not prepared to take that risk. But there seems to be no grounds for the consultants' definite expression of opinion that a policy based on the restraint of cars would be 'untenable' or 'inconsistent with the continuing prosperity of the city'.[17]

Some of the belief seems to have inspired, or perhaps was inspired by, the forecasting method that the consultants adopted, which was a variant of the transportation study method described in Chapter 3. Journeys made in Oxford in 1966 were surveyed and relationships established between the travel patterns found and the disposition of land uses, the transport facilities provided and the characteristics of the population, in particular car ownership. On the basis of the City's population and employment forecasts and land use plans, together with independently prepared estimates of future incomes and car ownerships, a projection of future travel by purpose, mode, origin and destination was made on the usual assumption that the relationships established in the analysis of the 1966 situation would continue to hold in the design year, 1991. According to this projection, the amount of road traffic in 1991 would have been nearly double that observed in 1966.[18]

The consultants seem to have been in two minds about how the figures resulting from this projection were to be interpreted. It was described as a trend prediction, a term which suggests that the figures represented what would be expected to happen naturally, in the absence of any unexpected circumstance or deliberate act of policy which would upset the trend. The following paragraph seems to have been written in that spirit:

This prediction of a 1991 travel pattern is based upon the extrapolation of current trends when applied to planning projections described in this Chapter. The pattern could be changed either by alterations in the projections or by any of the other methods of reducing travel demands, discussed in Chapter 6.[19]

On the other hand, the consultants recognised quite clearly that the projected levels of traffic could not come about unless roads were proved on an appropriate scale, and they pointed out that the necessary finance was highly unlikely to be forthcoming in the medium term (i.e. up to 1981 by when the bulk of the projected increase was expected) and that there was some doubt about it even in the long term.[20] Hence the practical difficulty was not to contain or alter the 'trend' but to find means of allowing it to continue.

Certainly, the projected figures were regarded as a desirable target, demands to be met if possible. They represented the achievement of the 'free choice situation'. The difference between the number of trips allocated to car according to the projection and the number that could be made by car if one of the designs being tested was adopted was taken as the degree of restraint implicit in that design.[21]

The objections to such an interpretation have been given in the last chapter. The 1991 trend figures were derived by projecting the 1966 observed travel pattern. Therefore the 1991 projections can only be taken as representing some kind of ideal or target for that year if the 1966 figures represented a situation which was ideal for 1966. In environmental terms, the 1966 situation was obviously very far from ideal. But even in transport terms no argument was given to show that the 1966 situation was ideal or even broadly desirable.

Even if that had been shown, it does not necessarily follow that the best policy for 1991 would be to strive to maintain the relationships observed in 1966. Such a policy is demonstrably expensive, and it would require to be shown that it was the best way of spending the money involved. This point applies particularly to a town that is expected to grow, as Oxford is, especially its business centre. As a town grows one would expect its transport system to rely more on public (or shared) means of transport as against private (or individual) means. Thus projections produced in the manner described can in an important sense be described as going against a natural trend.

None of this is intended as a particular criticism of the work done by the consultants; theirs was the approved approach and the weaknesses of it are only the weaknesses of orthodoxy. No doubt because it was orthodox it was accepted as the right approach both by Oxford City and by the Inspector at the 1970 Inquiry. The Inspector's opinion is expressed in the following passage from his report:

The need to do something to solve the traffic problem in Oxford has not been challenged. What is at issue is the means of doing it, and certain objectors advocate action to reduce the free use of private transport in the city centre and to increase the use of public transport, with an experiment in the first instance in the use of the existing road system only, including committed schemes, the traffic being distributed by traffic management measures and a park and ride scheme introduced.

It is accepted that the amount of restraint to be placed on private travel is a factor to be considered in the implementation of any scheme for Oxford, and any decision to reverse present trends would have to be supported by action to secure greater use of public transport. However, it is not considered that the objectors have shown there is no need for some form of new road construction to fulfil the objectives laid down in the terms of reference, particularly for a relief road around the central area. As regards the scale of construction, with the uncertainty attached to the question of future restraint, the policy of the council to plan on the basis of the trend growth of traffic would appear to be right. Certainly objectors have not demonstrated that it is not.[22]

In the Inquiry the City argued that the burden of proof lay with objectors to show that the City's plans were unnecessary; rather than with the City to produce any justification further to that provided by the consultants' study and this was the point of view adopted by the Inspector in his report. But it is not clear what arguments on the part of objectors would have been regarded as an adequate demonstration that the plans were unnecessary. Ultimately, the City's case rested on the fear that the city centre would not flourish unless access by car on the scale envisaged were provided. Since this fear was unsupported by evidence of experience either in Oxford or in other cities, the only way to test it would seem to be by experiment, as some objectors proposed. But the Inspector ruled out such an experiment as impractical:

To bring it into operation Magdalen Bridge would have to be closed to private traffic and a considerably increased public transport service would have to become available. It is uncertain whether the increased bus services would be available as a permanent scheme and it is most unlikely that the bus company would be prepared to increase their services for an experimental period.[23]

Since the City's proposals were admitted to be expensive, damaging and in some danger of not achieving some of their main objectives, it might be thought that the burden of proof lay with the City not with its critics. If experiments were necessary, the cost should be borne by the City and the Government, who would stand to gain, if the experiment were successful, by not having to spend money on road building. Perhaps the Inspector might have come to this view if he had not been influenced by the misleading policy principles implicit in transportation study methods.

The long story of Oxford's roads has therefore had an ending which is only partly happy. The original belief that it was wrong to exercise any control over the routes that drivers might choose, even when first class alternative roads had been provided, ruled out any chance of a satisfactory solution. The recognition that such control is entirely legitimate made it possible to locate the proposed new roads in positions which at least avoided doing irreparable damage to some of Oxford's most precious features. But the approved plan will still do damage enough: new roads and car parks of the kind proposed inevitably must. Seemingly attractive alternatives exist, but have not been seriously examined because of a continuing reluctance to admit that it is also legitimate to control modes of travel. There are many other historic towns which are likely to end up with second-best solutions unless that reluctance is overcome.

5. London: The Historical Background

London is so vast a city that no-one can hope to know it, and the same might be said of its planning history. To tell the full story even of the last few years would be quite beyond the scope of this book:* the aim is to tell enough of it to show how present plans have been inspired by the policy and design principles that have been described and why, in consequence, they will not succeed. The best way to do that is to concentrate on particular issues. Chapter 6 is concerned with plans for central London, as exemplified particularly in the current proposals for the redevelopment of Piccadilly Circus and Covent Garden, and Chapter 7 with the plan to build a system of motorways on a ring and radial pattern in inner and outer London. The present chapter sketches the historical background which is common to both these topics; finally, in Chapter 8, the main points of an alternative strategy are defined.

Road plans in London have a very long history, but for the present purpose it is sufficient to go back to the war years when Sir Patrick Abercrombie was preparing his *County of London Plan* and *Greater London Plan*, which were published in 1943 and 1944 respectively. Abercrombie attempted a comprehensive look at London's planning and transport problems. He borrowed much from his predecessors, but he was also quite prepared to alter their plans when he thought it right to do so, and anything he retained was in accord with his own conception of the problem.

The policy principles that prevailed at the time that Abercrombie was writing have been described in Chapter 1. It might be thought that London, because of its unique size, importance and history, would have been regarded as a special case where

---

* For a narrative history of London's road planning in this century, see *London Road Plans* 1900–1970.[1] For a more analytical account and appraisal see *Motorways in London*[2] and *Transport Strategy in London*.[3]

different principles would apply. But Abercrombie's principles are recognisably the same as those of the manual *The Design and Layout of Roads in Built-up Areas*. London, like any other town, had its own problems, but they were not such as to require a different approach.

As regards road traffic, Abercrombie saw congestion as the main problem. He believed that unless it was cured there would be a 'grave danger that the whole traffic system will, before long, be slowed down to an intolerable degree'.[4] The reason for congestion was that the number of vehicles had grown so fast that the road system was no longer adequate to cope with them.[5] The deficiences of the road system were of two kinds. There was no proper differentiation of roads according to their function; in particular, there was no distinction between roads for through traffic only and roads which afforded access. There were also not enough roads. Abercrombie's proposals were directed to remedying these two deficiencies. He believed that even in the central area, where he proposed much more extensive road building than anyone now suggests, the implementation of his proposals would solve the problem of congestion.[6] He did not contemplate the possibility of limiting the amount of traffic or giving priority to certain sorts of vehicles. The nearest he ever came to considering any such possibility was in the following reference to the central area:

With traffic increasing yearly, it may, indeed, ultimately become necessary to have a one-way system entirely and permanently, but we consider that such a contingency is sufficiently remote to make any prognostications mere speculation.[7]

Given his view that congestion could be cured by road building, it is not surprising that Abercrombie had nothing to say about the problem of buses. It is more surprising that he scarcely mentioned cyclists, except in a recreational context in Outer London, particularly since Tripp, to whose book Abercrombie contributed the foreword, had discussed their problems at some length. But Abercrombie did regard pedestrians as very important. One of the basic principles of the road plan was that pedestrians should be provided for adequately and safely. The differentiation of roads according to their function, and the system of precincts which he also recommended, were both intended to serve this end, but he also proposed more facilities for pedestrians, particularly in the central area.[8]

On matters of amenity and environment, Abercrombie's thinking is in advance of that of the manual. The manual emphasised the importance of such matters, but did not develop the theme. But the environment is central to Abercrombie's thought and proposals. He says quite plainly that: 'a revised plan, which allowed for a free and swifter movement of road traffic, would still be defective if the through traffic were allowed to invade the shopping street or to cut through the heart of a community'.[9] His criticism of many previous proposals for road construction was that they 'completely ignore the local conditions or civic pride of a particular district. Roads have been constructed or proposed with a first object of providing a straight route between two points and, if topographical conditions were satisfactory and some existing roads on the route could be utilised, scant notice was taken of the requirements of the local residents for quiet residential areas, with shopping and community centres separated from main roads and with some sense of homogeneity.'[10] In contrast to this, Abercrombie claimed for his own scheme that it was 'planned in direct connection with the existing communities and with proposals for development or continuance of the *status quo*. No civic or social centres are cut through or across, streets which in old days could serve the dual purpose of traffic and shopping have been sharply differentiated. . . .'[11]

In order to create this desirable separation of traffic from other activities, the first essential was to provide good arterial roads in well-sited locations away from town centres or other areas that it was wished to protect from traffic. The superior attraction of these 'by-passes' would draw away the through traffic, leaving in the by-passed areas, or precincts, only the traffic which had business there.[12] But Abercrombie did not rely on superior attraction alone: he saw that it would be necessary to close certain streets and to use traffic lights in such a way as positively to discourage the use of streets within the precinct by through traffic.

Abercrombie's policy for the protection of the environment was therefore simply the other side of the coin of his policy for providing for traffic and curing congestion. The extra roads that he proposed, combined with a sensible classification of roads, and some measures to ensure that the roads were used in the manner intended, were all that he believed was necessary to achieve both objectives.

Abercrombie had more to say about the roads than about the railways, and what he said has had a much more permanent influence, but that does not mean that he regarded the problems of the railways as less important or pressing. When he identified congestion as the main problem, he explicitly mentioned congestion on the railways as well as congestion on the roads.[13] It is clear that of all London's travel problems he regarded those of commuters as the worst.[14] He suggested a considerable number of improvements to the commuter railways, and he recommended that a special expert committee should examine these matters, which he found too technical to deal with adequately in his studies.[15] But the central feature of his plan was a massive outward movement of homes and work places, and it was to this above all that he looked for a solution to the problems of commuting and railway congestion, as is clear from the following passage from the *County of London Plan*.

As already indicated, there are in other parts of this Report certain important proposals which, relating to decentralisation and cognate matters, will affect the question of alterations in the railway system. It is considered that the solution of any existing railway problems is largely inherent in these proposals, and that no drastic alterations or additions to railways are necessary. When the large outward combined move of homes together with work is accomplished, the amount of daily travel between London and the suburbs will be materially reduced.[16]

Abercrombie did not see any conflict between his various proposals or any need to set priorities. For him, as for the authors of the manual, the question of cost could arise only in relation to means, not ends. He took the view that all these things would have to be done, somehow or other, well or badly; he argued that it would be more economical to do them well. He did not even attempt to work out the cost of his road plans or of any alternatives. The only alternative he discussed at all was a policy of widening all existing roads rather than building new ones. At one point, he seems prepared to admit that this might be cheaper, but he argues that it would be demonstrably inferior; elsewhere he maintains that it would probably not even be cheaper.[17] This attitude to costs is less easy to understand in London than in the smaller towns with which the Ministry of Transport's manual was concerned. In smaller towns, the cost of following the recommended policies might not have amounted to very much, but the implementation of Abercrombie's proposals for London would have required very great sums, as indeed he acknowledged. It was not long before his approach attracted criticism on this account.*

The general policy principles which Abercrombie held in common with his contemporaries, even together with his own analysis of London's particular problems, are not sufficient to explain why he supposed that the particular road network he suggested was the most appropriate, either in its general shape or in the scale of the particular roads. He inherited the general shape from his predecessors, and like his predecessors and contemporaries he was clearly very much influenced by considerations of geometry and symmetry. Tripp had declared that a ring and radial pattern was by far the best, and Abercrombie also saw it fitting in very well with the growth of London, which he analysed in terms of concentric rings. One function of Abercrombie's A ring was to provide a boundary to the central area,[19] and one function of the D ring was to mark the edge of the built-up area and the beginning of the Green Belt.[20] Why Abercrombie thought it desirable to mark these boundaries by an unmistakable physical feature such as a road is not clear, but he subscribed to the idea, more widely held then than it is now, that a city should be divided into zones containing a homogeneous or at least a predominant land use, and this may have had something to do with it. As well as fitting in well with the way that the city had grown, the road network that Abercrombie suggested seemed to fit naturally with the more decentralised pattern of land uses that he wished to see, and was perhaps also conducive to bringing about that pattern.

The fact that Abercrombie envisaged no control over the amount of traffic, even in the centre, is important to his whole

---

* For example, a paper 'The planning of ring roads, with special reference to London' read to a meeting of the Road Engineering Division of the Institution of Civil Engineers in December 1955 by Mr Rayfield had this to say. 'In other words, their proposals (i.e. the proposals of Abercrombie's *County of London Plan*) were not related to any specific period or to the money, material and labour resources available. To engineers, such an approach lacks realism.'[18]

ring and radial design. To some extent he viewed it as an efficient means of feeding traffic into the centre; he expected that the nearer one approached the centre, the heavier the flows would be. He placed the emphasis on the radials rather than on the rings: radials fed the centre and the function of the rings was primarily to link up the radials. This is clear from the following passage.

It is probably unnecessary to add that no ring is created for the purpose of a continuous gyratory traffic; it is to be viewed more in the light of the linking up of a series of cross-routes, connecting the radial roads. Another way of regarding the ring is a magnified by-pass, enabling traffic entering or leaving an area to avoid passing through and adding to the congestion of the centre. It follows that the amount of traffic on various sections of any one ring will vary with considerable abruptness, in contrast with that on the radials which tend to show a steady increase towards the centre as they are fed by the side streams entering them.[21]

Where Abercrombie's account is particularly thin is in his description of the type and volume of traffic that would use the roads he proposed. He usually described traffic only in terms of short and long distance or local and through traffic. He rarely described it in terms that would be standard today – by origin and destination, type of vehicle, purpose of journey, day of week and time of day, or in terms of the number of vehicles involved. This is not surprising, since he lacked the means to survey and describe existing movements, still more the means to predict future movements or to test the capacity of road networks to carry potential loads. Nevertheless, without such means it is difficult to see any way of justifying the claim that 'we put forward the road scheme with confidence as we believe it provides a basis for the satisfactory solution of London's traffic problems'.[22]

Abercrombie was perhaps most specific in his account of the way that his B ring, the ancestor of the present Ringway One, would be used. One of its functions was to take traffic out of the central area;[23] this justification which is still given, is discussed later in the chapter. Another important function was to carry traffic to the docks, and particularly to link the docks to the markets and industrial centres on the west of London.[24] One of the achievements of the London Traffic Survey was to show that this emphasis on docks traffic was much exaggerated, as indeed

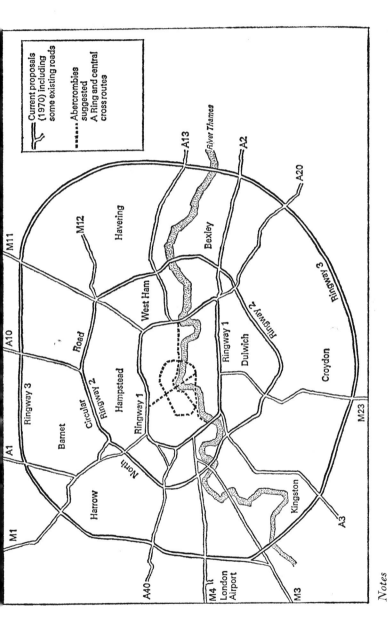

Legend:
———— Current proposals (1970) including some existing roads
········· Abercrombies suggested A Ring and central cross routes

Labels on map:
M11
M12
M1
A1
A10
Ringway 3
Barnet
North Circular Road
Ringway 2
Hampstead
Harrow
Ringway 1
West Ham
Havering
River Thames
A13
Bexley
A2
A20
Dulwich
Ringway 1
Ringway 2
Croydon
Ringway 3
M23
Kingston
A3
M3
M4
London Airport
A40

Notes

This map is diagrammatic only. Ringways 1, 2 and 3 correspond roughly to the B, C and D Rings proposed by Abercrombie. Ringway 4, corresponding roughly to Abercrombie's E Ring, lies entirely outside the GLC boundary and is not shown.

had been realised long before the Survey. What is left of the argument is destroyed by the current removal of the Port of London downstream to Tilbury.

Abercrombie's plan did not survive long in its full form. In May 1950, the Government indicated that it would not support the arterial 'A' ring-road, apparently on the grounds of the cost it would involve and the disruption it would cause.[25] The central cross-roads appeared only in a much reduced form in the LCC's Development Plan of 1951.

The central roads were crucial to Abercrombie's plans for Central London. Although he believed that congestion could be eliminated in Central London, he did not believe that it could be eliminated without the roads that he proposed. The drastic modification of these plans should therefore have led to a re-thinking not just of the road pattern of the centre but of the whole scheme for moving people and goods there, in other words, to an abandonment of the principle of the indiscriminate use of roads. It should also have led to a reappraisal of the road system for the rest of London, since Abercrombie conceived his design as a whole and in particular, as has been seen, saw it as a means of feeding traffic into the centre.

The development of movement plans for the centre and road plans for the whole of London will be discussed in the following sections, but before moving on to them it is of some interest to note that advocates of the GLC's* present motorway plans still appeal for support to Abercrombie's judgement: it is claimed that his experience and intuition led him to appreciate broadly what should be done even though he lacked the advantage of modern analytical techniques. Two quotations will illustrate this point.

A point needs to be made regarding the *pattern* of the primary network. It is perfectly true to say that the essential pattern was settled long before the London Traffic Survey was started. In fact the pattern is quite plainly a fairly direct derivative of the proposals advanced in the County of London Plan and the Greater London

---

* The Greater London Council succeeded the London County Council on 1 April 1965, but is responsible for a much wider area. The LCC's boundary was roughly that of Inner London or Victorian London. The GLC's boundary is roughly that of the continuous built-up area of London.

Plan of 1943 and 1944 respectively. . . . The fact seems to be that Abercrombie, although unable to quantify and predict in the way that is possible with the help of today's analytical tools, had a very good instinct as to the pattern of roads required. One matter which he saw very clearly and which has been validated by all subsequent studies, was the need to counteract the strongly radial pattern of the London road system by the insertion of orbital or tangential routes. . . . Indeed it might be thought that an extraordinary feature of London planning is the way in which the concept of a ring and radial road system has survived for many years even though practically nothing has been built. We ourselves have certainly seen no reason to challenge the basic pattern at this late stage.[26]

Abercrombie propounded the relative simplicity, need and logical inevitability of developing a ring/radial pattern of major routes in London. The Ministers of Transport and of Town and Country Planning, in a joint statement in 1947, approved the Abercrombie road strategy. . . .

Since then, extensive surveys, notably the London Traffic Survey, have been carried out on the demand for traffic movement in the Greater London area and these have provided a quantitative basis for the planned development of trunk roads in and around Greater London.[27]

It is certainly true that Abercrombie's description of the defects of the workings of London's transport system is both comprehensive and enlightened, more so than other more recent appraisals have been. But his positive proposals were, nevertheless, based on views which were either incorrect or at least have been overtaken by events. The massive decentralisation of jobs and homes which he planned for has not taken place, neither has the easing of the commuter and railway problems which he expected, naïvely perhaps, would follow from decentralisation. His belief that road building could solve congestion was mistaken, and in any case the roads now proposed, although obviously derived from his proposals, differ from them in a crucial respect. Finally, as we shall see, the account now given of the use which it is expected will be made of the roads is quite different from anything in Abercrombie's analysis. It is therefore incorrect to suggest that Abercrombie produced reasons for the motorway proposals which later more scientific work has confirmed,

6. Central London

Although Abercrombie's proposals were considerably modified in the LCC's 1951 Development Plan, that Plan contained extensive plans for road building in Central London. The amount of road building that took place in subsequent years, in the centre as elsewhere in London, fell far short of what had been proposed and approved, apparently because of financial restrictions. But a certain number of major road widening or building schemes were undertaken, usually with the purpose of increasing the capacity of particularly congested points such as Park Lane, Hyde Park Corner and Knightsbridge.

Two other important developments were the introduction of on-street parking control and of traffic management measures. Parking meters were first installed on an experimental basis in the West End in 1958 and within a few years their use had spread widely over Central London. They were originally seen as a means of controlling indiscriminate parking which hampered traffic flow and it was stipulated that any surplus revenue should go to the provision of off-street car parks. At about the same time, traffic management measures were introduced. They, too, were seen as an aid to traffic problems, narrowly conceived in terms of enabling the road system to accommodate as many vehicles as possible. This is clear from the following passage taken from the Ministry of Transport's Annual Report for the year ended 31 March 1964:

THE RESULTS OF TRAFFIC MANAGEMENT IN LONDON
It is now four years since the London Traffic Management Unit was set up in the Ministry of Transport. In these four years traffic management techniques have been applied extensively and systematically. What were at first innovations have become commonplace and all the time, traffic conditions in London have been improving. The improvement has been gradual, and so may run the risk of passing unnoticed. But the benefits that Londoners have been drawing from these policies

are both real and substantial. Figure 4 (page 49) shows how traffic conditions on some 40 miles of main streets in Central London have changed in 10 years.

Figure 4 is derived from surveys which the Ministry of Transport and the Road Research Laboratory have carried out between 1952 and 1964. The surveys were made in both off-peak (9.30 am to 5 pm) and peak (5 pm to 6 pm) traffic conditions on weekdays over a period of four weeks in each of the years shown. In the eighteen months between October 1962 and April 1964, traffic volumes increased by 5% in the off-peak and $7\frac{1}{2}$% in the evening-peak periods: in spite of these increases, overall journey speeds went up by 13% off-peak and 9% in the evening-peak. In the $3\frac{1}{2}$ years from October 1960 to April 1964, the increase in volume was 10% off-peak (15% in the peak period) and the increase in speed throughout the day was about 20%. [The diagram referred to shows speeds and flows on Central London streets. The term 'overall journey speeds' does not refer to door-to-door speeds].[1]

It should perhaps be mentioned that the success of these schemes, even when judged in fairly narrow traffic terms, has been questioned,[2] but in addition their implementation implied the adoption of the arbitrary and unfortunate order of priorities that has been discussed in Chapter 2.

In spite of the great benefits claimed for traffic management, it was not supposed that traffic management was a substitute for road building. However, in the early 1960's the objective of trying to accommodate all the traffic that could be foreseen was modified with respect to Central and Inner London. This modification became generally known during the long controversy over the redevelopment of Piccadilly Circus. The controversy started with plans for developing a particular site adjacent to the Circus, but its scope widened, both in terms of the area covered and the considerations involved. It was conducted with unusual openness and at a high political level and therefore provides a unique insight into official thinking at the time.

Piccadilly Circus

In 1959, a developer who had acquired one of the important sites bordering Piccadilly Circus, obtained provisional planning permission to put up a building, which included a tower 172 feet high that was to form the background for illuminated advertising. The developer was unwise enough to give a press conference at which his plans were illustrated. They aroused a storm of protest, particularly in the architectural papers and among the architectural profession, both on the grounds of the scale of the proposed building, which would have been visible from many parts of London and beyond, and because architecturally it was extremely undistinguished. This protest led the Minister of Housing and Local Government to call in the application and set up a special Public Inquiry into it. The Inspector (Sir Colin Buchanan) reported unfavourably on the scheme and the Minister consequently refused permission for it, and when doing so asked the LCC to prepare a comprehensive plan as a guide to the redevelopment of the Circus as a whole.

The LCC commissioned Sir William Holford (now Lord Holford) to prepare such a plan, and the report describing his scheme was published in March 1962. Holford's scheme would have allowed for a 20% increase in the volume of traffic that had used the Circus in 1960, an increase which was judged insufficient by the Ministers of Housing and Local Government and of Transport, who therefore rejected the scheme. The view of the Ministers was given in a letter to the LCC dated 2 September 1963, in which they said that the proposals were 'now seen to do too little to provide for the almost inevitable future traffic increase; and were bound to result in, and perpetuate an inadequate traffic provision at a key junction in the centre of London'.[3] Expanding this argument, the letter said:

All the evidence becoming available to the Minister of Transport from the Hall and other reports, makes it clear that an even bigger increase of traffic in Central London than had been anticipated must be expected over the next two decades. Much of the information about the growing weight and significance of the urban traffic problem has, of course, become available since the Holford plan was prepared.

In the Circus itself there is little doubt that the traffic flow will increase to a substantial extent. In 1960 – the year in which Sir William Holford was appointed to prepare his scheme – 56,359 vehicles passed through the Circus in the twelve daylight hours. In 1962 the comparable figure was 62, 109; an increase of 10%. The road improvement scheme at the junction of Knightsbridge and Sloane Street, the widening of Knightsbridge, the eventual widening of

Piccadilly by the Ritz Hotel, plus possible improvements to Charing Cross Road and other like schemes, will make additional traffic in the Circus inevitable, particularly through the critical north-west section.

The traffic needs of the Circus should, it is suggested, be related to the recommendations of the Minister of Transport's Design Working Party. The Working Party proposed that the reserve capacity of any improvement in Inner London should either allow for a 60% increase on the volume of traffic at the date of completion, or the maximum foreseeable increase on the approach roads, whichever is smaller. In the case of Piccadilly Circus it is considered that the capacity of the approach roads – the decisive factor here – will be limited to about 50% above 1960 figures and that, therefore, this is the best criterion to apply. This would mean about 85,000 vehicles for the twelve-hour period.[4]

It is a mark of how confused thought on this subject had become that a considerable increase in traffic should have been simultaneously described as inevitable and as something that would not come about unless the existing plans were drastically modified.

Lord Holford reported to the LCC that he could find no way of securing the suggested traffic capacity while at the same time retaining Piccadilly Circus as a place of attraction and resort and suggested that 'other means should be found, probably outside the Circus itself, of regulating its motor traffic to some degree'.[5] The LCC concurred with Holford's view. After discussions between the LCC and the Ministries, a working party was set up with the following terms of reference:

To determine the area which is of significance in relation to the traffic passing through Piccadilly Circus, and to consider probable developments in that area affecting the volume and composition of that traffic in the foreseeable future; to consider what measures could be taken in that area during the next twenty years to deal with the traffic expected; and in the light of this to assess the load of traffic for which the Circus will have to provide.[6]

The membership of the working party was drawn from high-ranking staff of the Ministry of Housing and Local Government, the Ministry of Transport, the LCC, Westminster City Council and the firm of consultants then engaged on the London Traffic Survey. In setting about their task, they made it clear that they

Figure 10   Central London

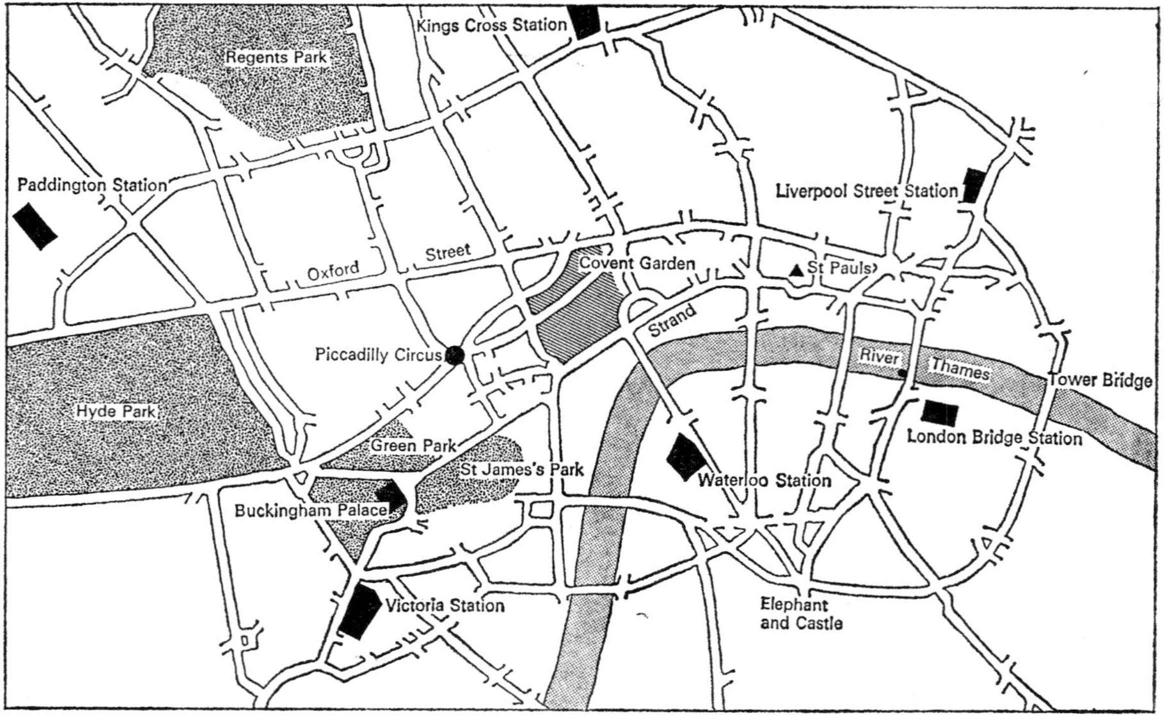

Regents Park

Paddington Station

Kings Cross Station

Oxford Street

Hyde Park

Piccadilly Circus

Green Park

St James's Park

Buckingham Palace

Victoria Station

Covent Garden

Strand

St Pauls

Liverpool Street Station

River Thames

Tower Bridge

London Bridge Station

Waterloo Station

Elephant and Castle

were concerned at least as much with environmental considerations in Piccadilly Circus and the surrounding area as with traffic. They were also concerned, in achieving a reasonable balance between traffic and environment, to rely on 'methods that are within our means'.[7]

The working party first considered whether it was right to plan for an increase of 50% in traffic in the general area of the Circus, as had been specified in the Ministers' letter. After conducting 'a critical re-examination of the future trend of traffic demand in the Circus' and considering various measures for 'regulating the potential demand' they concluded that an increase of that order would have to be allowed for in the general area. Ways of keeping the traffic away from Piccadilly Circus itself were examined, ranging from a potential new network of primary distributor roads in the centre of London to more local relief roads. But it was concluded that such measures would be inpracticable, ineffective or unacceptable environmentally. Hence it followed that a 50% increase of capacity would have to be found within the Circus itself. The only acceptable way of providing the extra capacity was to give over the ground level to traffic and to create a new level for pedestrians above the traffic. This new level was not to be simply a platform at the Circus itself; it was to be extended, sooner or later, into some of the neighbouring streets and would provide access to the adjoining buildings.[8]

To have increased the capacity of the junction at Piccadilly Circus by 50% would not in itself have permitted a comparable increase in the volume of traffic through the Circus, since the capacity of each of the approach routes was limited by various bottlenecks or 'pinch points', which mainly occurred at other road intersections. But on examination the working party concluded that a combination of measures would make it possible to achieve the required extra capacity on the approach roads. The necessary measures were described as follows:

Various combinations of methods are possible, including greater emphasis either on traffic management (one-way routings, etc) or on road improvements, though both would be required. The essential components would be major road improvements to overcome the 'pinch points' (not necessarily at the junctions themselves), together with the following measures at various points of the system:

(i) extension of linked traffic light system;

(ii) removal of parking meters to free more road space for traffic and to ease traffic flows;
(iii) strict enforcement of 'no waiting' restrictions;
(iv) resiting of bus stops clear of signal junctions;
(v) banning of right turns at signal junctions;
(vi) other traffic management measures, including extension of one-way working.[9]

There is little doubt that the various proposals of the report would suffice to bring about the desired 50% increase in traffic capacity; it is less clear that they are compatible with other desirable objectives, including those that the working party recognised. In arguing that their proposals would produce an acceptable environment, the working party claimed support from the ideas of traffic architecture that had been propounded in *Traffic in Towns*.

In short, there exists here the opportunity for an outstanding achievement in what the Buchanan report calls 'traffic architecture'. The following quotation from that report is singularly pertinent in the context of the Circus and its redevelopment:

> 119. Although traffic architecture techniques would involve a 'new look' for urban areas, in many ways it could still result in an 'old look' freed from the domination of the motor vehicle. To take an extreme but simplified case, the central area of a town might be redeveloped with traffic at ground level underneath a 'building deck'. This deck would, in effect, comprise a new ground level, and upon it the buildings would rise in a pattern related to but not dictated by the traffic below. On the deck it would be possible to recreate, in an even better form, the things that have delighted man for generations in towns – the snug, close, varied atmosphere, the narrow alleys, the contrasting open squares, the effects of light and shade, and the fountains and the sculpture.[10]

But the proposals of the Piccadilly Circus report could not conceivably realise the ideal presented in this passage.* In the first

---

* Even the most ambitious of the schemes illustrated in *Traffic in Towns* itself, which did consider the complete rebuilding of a central area at an upper level, does not come near to achieving this ideal. In spite of complete physical separation, pedestrians are never free from the consciousness of the presence of great numbers of motor vehicles and are confined to a rigid pedestrian network, rather than being able to flow freely over the entire open space between buildings.[11]

place, they are much too limited in geographical extent. In saying that an increase in demand of 50% had to be allowed for, the working party relied on forecasts from the London Traffic Survey, which purported to show that an increase of this order could be expected in the West End as a whole. But it was not proposed to double-deck the entire West End, and the working party dismissed, for the foreseeable future, the idea of applying Buchanan's concept of a distributor network and associated environmental areas to Central London. Its positive suggestions for the West End were vague and not reassuring.

The fact that full realisation of the network system and environmental area concept lies far in the future does not mean that the traffic problem is insoluble. Professor Buchanan has himself stressed that, if the ideal solution of the primary network is unlikely to be achieved within the foreseeable future, then not only will there be a much lower limit on the amount of traffic that can be accommodated but other means of reconciling traffic and environment must be found.

One means of achieving at least a measure of reconciliation lies in the manner in which areas that are due for redevelopment are in fact redeveloped. The opportunities for urban renewal on a massive scale and on an entirely new pattern are only now beginning to emerge.

As redevelopment proceeds it should be possible to establish at least the rudiments of an 'embryo network', based largely on existing streets, and to attract to these improved routes traffic which at present filters through the potential environmental areas. Traffic management, coupled with limited road improvements, and the opportunities arising in conjunction with major redevelopment, should enable progress to be made with a positive policy of this kind over the next ten to twenty years.[12]

In the more limited area surrounding the Circus, the measures listed above which were deemed necessary to achieve the required extra capacity on the approach roads are clearly not conducive to a policy of restoring civilised conditions. At the Circus itself, and in the immediately surrounding streets where an upper level pedestrian network would be provided, pedestrian safety would be achieved, but not necessarily pedestrian convenience, and certainly not freedom from other kinds of traffic intrusion. The report also recognised that the architectural problems involved were formidable, and gave no satisfactory reason to suppose that they would be solved.

In spite of their warning to keep within our means, the working party gave no indication of how much money might be made available and said very little about what the proposals would cost. At the Circus itself, it was estimated that the public works costs would be of the order of £5 or 6 million, but no estimate was made of the costs of acquiring property except that they would 'be heavy, and similar to those involved in the Holford scheme'. The cost of extending the upper level pedestrian piazza from the Circus into the adjoining areas was not estimated, although some at least of the extensions were regarded as essential to the success of the original scheme. It was said that some of the measures necessary to increase the capacity of the approach roads would be 'very expensive', but no more precise indication of the cost was given, nor was anything said about the cost of the long term improvements hoped for in the rest of the West End.[13]

The aim of providing 50% more traffic capacity in this part of London is so difficult to reconcile with other desirable objectives that it is important to understand more fully why the working party decided to adopt it. A special survey was commissioned to ascertain how much of the existing traffic was essential. This survey showed in the first place that only about a quarter of the traffic in the Circus was going right through the West End, which suggested there was little to be gained from attempts to divert the traffic.[14]

This left the crucial question of whether all the journeys had to be made by the same means as were in fact employed: a particularly pertinent question since the survey showed that 70% of the vehicles in the Circus were vehicles providing individual means of personal transport – motorcycles, taxis and cars.[15] The working party did not tackle this problem explicitly. It was pointed out that commuters who did not use their cars for other purposes during the day and people on shopping trips accounted for only a small percentage of the traffic; all the other traffic was implicitly treated as essential.[16] The great volume of taxi trips was noted, but the only comment made was that 'the availability of this service reduces the amount of private car travel that would otherwise occur and is a useful alternative means of catering for this type of journey'.[17] The possibility that buses might cater for some of the journeys made by car and taxi was not discussed, neither was the Underground ever mentioned as a possible competitor.

To estimate future requirements, the working party first examined the trend in the number of vehicles using the Circus, as revealed by biennial police censuses. These figures showed that there had been no increase, and perhaps some slight decline, over the preceding years in total traffic going through the Circus, but there was some evidence that one-way traffic schemes had brought about an increase in the traffic entering it from the south and west, which the working party judged to be a critical consideration. The working party also examined the first set of forecasts then becoming available from the London Traffic Survey. It was predicted that the number of journeys made to the West End by car and motorcycle would nearly double, as between 1962 and 1981, and that the amount of traffic of all kinds in the West End would increase by 40% or 50%. These forecasts were treated as a forecast of demand and hence presumably of need.[18]

The working party apparently believed that the increase in traffic capacity which they proposed would produce some kind of final solution to the problems of the Circus. They were very much alive to the phenomenon of traffic generation, recognising that: 'The general tendency in London is for any additional traffic capacity provided by road improvements or traffic management measures to be fully absorbed by additional traffic flow within a relatively short period – subject only to the limitations imposed by bottle necks elsewhere on the system.'[19] But they seem to have believed that an increase of capacity of the order they were considering would be sufficient to overcome this tendency. Hence they apparently believed that under their scheme traffic would no longer be restricted by congestion, which they criticised as 'an entirely indiscriminate means of control and a highly uneconomic one', and after further criticisms went on to say: 'For these reasons we must reject as unrealistic and uneconomic the proposition that the traffic problem of the Circus can be solved, or at least mitigated, by retaining the present points of congestion that limit the capacity of the approach routes.'[20]

The reason why the working party believed that their proposals would be adequate to cope with generated traffic was

D

presumably that they supposed that the London Traffic Survey's forecasts took account of the suppressed demand that the provision of more traffic capacity would release. If this was the working party's view, it was certainly incorrect. In 1962, when the basic traffic surveys were carried out, there was very great suppressed traffic demand in London, and the methods used to project the 1962 situation to 1981 implicitly assumed that there would be a similar degree of suppressed demand in the future. In any case, the London Traffic Survey forecasts were only to the year 1981, by which time it would scarcely have been possible to complete the proposed works. The same methods applied to a later year would have produced higher forecasts.

The working party's report was published in the spring of 1965. It was well received by the Ministers concerned and was the basis of some more detailed designs which were then put in hand. These in turn were officially approved early in 1967 and apparently remain the foundation of present plans for the redevelopment of Piccadilly Circus, Regent Street and other streets. In the meantime, however, the objective of accommodating a substantial increase in traffic in Central London has been abandoned, as will be seen below.

Covent Garden

The most ambitious and important scheme now proposed in Central London is the plan for redeveloping Covent Garden. The basic approach to traffic problems in this plan is the same as in the plan for Piccadilly Circus, but the area is larger and very different in character, so the proposals throw more light on the problems of applying this approach to central areas of large cities.

The stimulus for the Covent Garden plan is best described in the opening words of the official document.

In 1972, after three centuries on its present site, the Covent Garden market is expected to vacate the fifteen acres it now occupies in the heart of London. Taken together with a bigger adjoining area containing much property ready for redevelopment, this will provide the opportunity within the next ten to twenty years, to reconstruct up to eighty acres of the West End – an area big enough to allow breaking away from the existing urban pattern and building in a new form better adapted to modern needs. The pattern of streets and buildings has changed little from that shown in drawings dated more than 200 years ago – a pattern based on the ancient route of the Strand and extended haphazardly outwards from Inigo Jones' plans of the early 1600's for the Covent Garden 'Piazzas', the first of the London squares. It is not surprising that the pattern itself is obsolete. What is unusual is the scale of the present opportunity to make basic improvements.[21]

In order to achieve the desired reconstruction, a Covent Garden planning team was set up in 1965 under the auspices of the local authorities concerned and was given the following objectives:

(a) The incorporation of a complex of uses to create a vigorous and interesting environment by day and by night both as a place to live and as a centre for entertainment and cultural activities;
(b) A substantial increase in residential accommodation;
(c) The provision of new public open space in addition to amenity open space within individual sites;
(d) The easing of congestion in Central London, in particular by the avoidance of major employment generators and major traffic generators;
(e) Separate but integrated systems for pedestrian and vehicular movement within and immediately adjoining the area, including specifically proposals for efficient coordination with public transport and for car parks on a scale to be recommended based on a study of traffic generation following redevelopment and of the capacity of the approach roads;
(f) The integration of new development with existing uses and some provision for the retention of suitable mixed uses which are appropriate for the area's special location and character;
(g) The retention of those groups of buildings, including buildings of architectural and historic importance, which contribute substantially to the variety and character of the area and are the physical embodiment of its past history.[22]

It was also made clear that the team should bear in mind the desirability of excluding through traffic[23] and should keep a close eye on costs, on which it was said:

In formulating its proposals, the Team will have to balance on the one hand the importance of the site and the scale of expenditure

Figure 11   Covent Garden

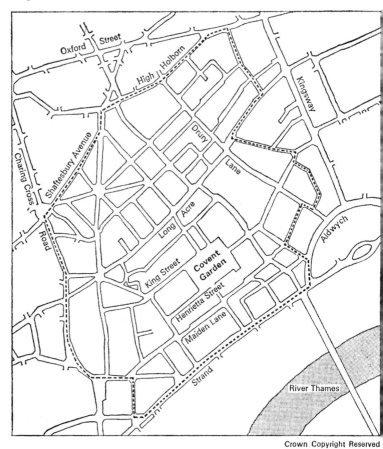

necessarily involved, and on the other, the need for economical solutions and the accommodation of remunerative uses to the maximum compatible with the basic objectives. The heart of London may deserve special expenditure, but the great demands on public funds necessitate strong efforts to minimise the public cost of redevelopment.[24]

The plan was prepared in two stages. In 1968, a draft plan *Covent Garden's Moving* was published for public information and comment. The revised plan *Covent Garden, The Next Step* was published in May 1971. The revised plan contained a number of important alterations of detail, but the broad conception was unchanged.

One of the most important features of the plan is the proposal to provide more capacity on the main roads on the perimeter of the Covent Garden area. In the draft plan, it was intended to treat every main road in this way,[25] but this was modified in the revised plan. The most important such proposal, which survives in the revised plan, affects the south side, where it is intended to make the Strand into a one-way street running west, and to provide a new four-lane eastbound road within Covent Garden itself, along the line of Maiden Lane. The original plan to widen the roads on the west side, Charing Cross Road and Shaftesbury Avenue, also still stands.

Within the area, a drastic reconstruction of the existing road pedestrian networks is proposed. This would not be on the lines of the Piccadilly Circus scheme, with complete vertical segregation of traffic and pedestrians; pedestrians would at some points be on a new upper level, but at other points would make use of the existing streets, of which some would be reserved exclusively for pedestrians and others would continue to carry vehicles as well.[26] It is also proposed to provide 4,000 off-street parking spaces, in addition to the 600 off-street parking places already existing within the area.[27]

Very considerable redevelopment of the buildings in the area is envisaged. In terms of floorspace, less than 40% of what is now in the area would remain. The redeveloped area would contain 20% more floor space than the present area, allocated to rather different uses. In particular, hotels, which account for a negligible proportion of the present floor space, would comprise more than 10% of the proposed floor space, and a big new international conference centre is proposed. But there would be a substantial decline in the floor space devoted to industry and commerce.[28]

To what extent would the proposals achieve the objectives originally specified? This is not an easy question to answer, since the objectives are not always set out in a way which makes it very easy to tell whether they have been achieved or not. There will certainly be an increase in the floor space devoted to residential accommodation. But whether or not this is sufficient in relation to the contribution that this area could make to the relief of London's housing problem is another matter; it is clear that the rents will be too high for some of the present residents to be able to remain in the area.[29] It also appears that new public open space has not been provided on the scale suggested by the application of normal GLC standards; on the other hand, it is argued that it would be inappropriate to apply such standards to the central area, and that the pedestrian spaces provided should be thought of as a substitute for conventional open space.[30] Whatever the merits of this argument, it does not seem to meet the objection that none of the open spaces will contain play facilities for children, which surveys show to be a particularly valued facility.[31] This omission seems particularly unfortunate in the light of the aim to increase residential accommodation in the area.

What does seem clear is that there has been only a limited success in achieving the transport objectives of the plan. According to the official figures, the redeveloped Covent Garden will generate about as much traffic as the area generates now, even though a substantial part of the present traffic consists of market vehicles, which will disappear when the market moves;[32] hence it cannot be claimed that the plan will contribute to the 'easing of congestion in Central London'. Complete pedestrian separation has not been achieved, and car parking has not been related to traffic generation as laid down in the brief.[33]

Graver doubts still surround the question of whether the character of the area would be retained, either in physical terms or in terms of the agreeable mixture of uses which the area now contains. It would be difficult to demolish 60% of the buildings in an area and to change its layout radically without altering its character. Even if the 40% remaining included all the individual buildings of merit, the character of an area such as Covent Garden is largely formed by the presence of buildings which are individually unremarkable but form an agreeable and harmonious whole. Also, there is no doubt that the 60% of buildings to be demolished contain many fine buildings, particularly in the south of the area. This is indeed admitted in the official documents; for example, it is stated that 'King Street and Henrietta Street are of considerable interest architecturally but it is doubtful if the minority of good buildings could be preserved in the face of large scale redevelopment plans'.[34] The official documents also give grounds for doubting whether the existing mix of uses can continue, and particularly whether it would still be possible to accommodate in the area activities which require both a central location and low rents. It seems likely that the area would become increasingly like the rest of the West End, and would lose its individual flavour.[35]

The capital costs of redevelopment were estimated in the draft plan at £141 million,[36] although newspaper reports suggest that the revised estimates are a little higher.[37] The greater part of the sum involved is to be found by private developers. Although the planning team was enjoined to bear in mind the need to find economical solutions to the extent compatible with the achievement of the basic objectives set, this term in their brief is so loosely worded that it is impossible to tell what level of expenditure would be consistent with it. What is clear is that the proposals of the draft plan would have involved a net cost to the public authorities, which was claimed to be 'reasonable relative to the public improvements proposed', although 'it would nevertheless involve a charge on the rates, after allowing for Government grants'.[38] It is not clear from the documents to what extent the alterations in the revised plan have reduced the net public cost of the scheme, but at best it appears to be still unprofitable.

There is therefore a serious doubt, to say the least, whether the Covent Garden proposals will succeed in their objectives. It is the transport proposals that cause the trouble.

The aims of preserving the character of an area and at the same time radically changing its road structure and urban form are virtually contradictory. Such a drastic restructuring must also be expensive, not simply because roads are themselves expensive items, but because so many buildings have to be replaced which might have been left – for example, the plans involve moving two theatres[39] – and other buildings which might have been provided cheaply must cost more in order to conform with the multi-level design proposed. The cost in turn forced a change in the land use plan; in order to recover some of the expense, the planning team was obliged to add to the total amount of floor space that it was originally intended to provide, and to give over a greater proportion of the floor space to uses that could command high rents.[40] In this way, it seems that the cost defeated the aim of keeping down traffic generation, since the more intensive development proposed required more movement and more vehicles in order to support it. This is indeed the final paradox of the plan, that a design adopted primarily in order to produce a civilised environment relatively free of traffic should in fact have led to the generation of more traffic.

It is important to understand the reasons for the adoption of such troublesome road proposals. Two reasons were given for increasing the capacity of the peripheral roads: to provide for the extra volume of traffic in Central London generally and to cater for the through traffic which it was intended to displace from Covent Garden itself. The exclusion of through traffic was also one of the reasons for the proposed internal road system, which is so arranged as to make it impossible for through traffic to go through the centre. The other reason for the proposed internal road system was to provide the separate pedestrian network specified in the brief.

The need to provide more traffic capacity in Central London as a whole is mentioned several times in the draft plan, *Covent Garden's Moving*, and appears to be the more important of the two reasons given for increasing the capacity of the peripheral roads, and in particular for duplicating the Strand.[41] But by 1971 the unwisdom of this objective had been widely recognised; it is not mentioned at all in the revised plan and only in the most shadowy form in the documents produced for the Inquiry held in the summer of 1971. As a partial substitute, some mention is made of the desirability of serving existing traffic better and making some allowance for reserve capacity.[42] However this considerable change of view did not lead to a corresponding change in the plan; the proposals for the Strand, Charing Cross Road and Shaftesbury Avenue were virtually unaltered, and even the changes on the northern and eastern perimeter roads seem to have been made for other reasons.*

Covent Garden attracts very little through traffic at the moment, because the layout of the streets and the presence of stationary market vehicles combine to deter it.[44] So if all the

* The desired widening of High Holborn had already been achieved; of Kingsway it was said that the need for the relief road originally envisaged was considered 'unlikely to arise within the time period covered by the Covent Garden plan'.[43]

through traffic were transferred to the peripheral roads, the extra capacity envisaged would appear to be excessive. Moreover, the argument depends on the view that the through traffic constitutes a fixed load which, if excluded from Covent Garden would add an equivalent amount to the flow on the other streets. If the corresponding extra capacity on the peripheral roads were not provided, this would not happen; there would be a downward adjustment of the traffic on the central area network to allow for its reduced capacity. But even if it could be demonstrated that the present level of traffic in Central London was desirable, so that the capacity of the road network ought not to be reduced, it still requires to be shown that it is preferable for the traffic to run on the peripheral streets rather than through Covent Garden. The peripheral streets, particularly on the south and west, can by no means be regarded as mere distributors with no other function to perform but to carry moving traffic. Covent Garden is an extreme example of the possibility that the cure of removing through traffic may be worse than the disease of allowing it to remain.

This becomes a more serious possibility still if part of the cure is the provision of the elaborate internal network proposed. This would certainly be an effective way of keeping out through traffic, but it would create longer journeys for many kinds of local traffic, besides contributing to the considerable difficulties of conserving the area that have been described. Once it is allowed that some through traffic is permissible, and may well be preferable to throwing extra loads on the peripheral roads, it is easy to suggest ways in which the volume of through traffic on the present street system could be prevented from rising to excessive amounts when the restraining influence of the market vehicles is removed. Appropriate measures might be to impose very strict speed limits, to give the roads corrugated surfaces at certain points, and to use traffic lights to impose delays at critical junctions.

But the more important reason for advocating the internal road network as proposed was undoubtedly the feeling that the existing street form was out of date in modern conditions, and did not allow for the reconciliation of the needs of vehicles and pedestrians. It is important to note that this view was put forward before any study of the area had been undertaken; as we have seen, the statement that the street pattern was obsolete appears in the very first paragraph of *Covent Garden's Moving*.

Similarly, when the brief refers to pedestrians, it does not specify the ends to be achieved – the provision of safe, convenient and agreeable conditions for pedestrians – but the means, the provision of 'separate but integrated systems for pedestrian and vehicular movement'. It was assumed, with no further inspection or argument, both that existing conditions were not tolerable, and that the means specified were the only possible way of achieving the desired improvement.

The irony of this is that the present conditions for pedestrians in Covent Garden are exceptionally agreeable, and it is far from clear that what is proposed would be any more convenient. At present, there are very few completely pedestrian streets, and pedestrians are sometimes obliged to pick their way between standing vehicles, but, because of the standing vehicles and the intricate street pattern, it is impossible for vehicles to move at all fast or to dominate pedestrians. Hence it is possible to traverse the area on foot in any direction in harmonious surroundings and with few interruptions. If these conditions are achieved, the physical separation of vehicles and pedestrians is of small importance. If the proposals were implemented, the desired separation would not be completely achieved, there would be several changes in pedestrian level, access to neighbouring buildings would not always be possible from the pedestrian network, and many parts of the pedestrian network would be subject to the noise, fumes and visual intrusion of motor vehicles.

If it were true that the present layout of Covent Garden was obsolete and inherently unsuitable for modern needs, the same would be true of all our old towns and of most of Central London, to which exactly the same arguments apply. It is clear that the authors of *Covent Garden's Moving* did in fact believe that the street layout of Central London was seriously deficient.[45] This belief, which is equally evident in the Piccadilly Circus report, and is presumably largely inspired by *Traffic in Towns*, no doubt explains why it was thought possible to write the requirement of vehicle and pedestrian separation into the brief in advance of a study of the particular problems of Covent Garden. It is clear from *Covent Garden's Moving* that the reason why the present pattern of streets and pavements was thought to be out-of-date is because of the actual and expected rise of traffic: 'The phenomenal rise in the volume of traffic and its increasing speed due to road improvements and traffic management schemes, makes more urgent the need to separate people from traffic.'[46] And again: 'In modern cities, the great increase in motor traffic has enormously increased the disadvantages of the conventional street pattern in which vehicles and pedestrians share the same routes and, as a result, are frequently in conflict.'[47] Once again, the increase in motor traffic is treated as being inevitable or at least as unquestionably desirable, even though one of the terms of the brief was to ease congestion and avoid traffic generators.

Many of the individual buildings in Covent Garden are obsolete and due for replacement or substantial renovation. But once the mistaken view that more and more vehicles must be accommodated is discarded, there is no reason at all to accept the basic presupposition of the plan that the whole street pattern is out of date. The pattern was formed in an age when pedestrian movement in cities was even more important than it is now, and it was, and remains, a very convenient layout for pedestrians. All the other important objectives of the plan would be as easy or easier to achieve within the framework of the present street system than under the proposed arrangements: the conservation of the physical character of the area, the retention of diverse uses within it, including those which cannot command high rents, the avoidance of traffic generation and the saving of public money.

Present Policies for Central London

It has been mentioned that it is no longer official policy to attempt to accommodate substantial traffic increases in Central London. To have given up this objective represents a major reversal of policy, but it was not done in a clear and explicit way, so that it is very hard to point to a moment when the policy was changed or to identify the precise reasons for the change. However, the cost and other difficulties of pursuing the former policy have become steadily more apparent over the years, and so has its pointlessness. The vast majority of people come to Central London and move about within it by public transport, and even massive increases in road capacity could change this pattern so little that there seems to be no point in trying. Table 7 illustrates these points, and also shows that there has been no increase, but even some decline, in the number of people coming to Central London over the years. These appear to be the considerations that have brought about the change in official policy.

The Written Statement of the Greater London Development Plan, published as recently as 1969, does not in fact contain a

TABLE 7

Daily arrivals in Central London between 0700 and 1000 hours
Units: thousand travellers

Year	BR or Underground	Bus and Coach	Car	Motor Cycle	Pedal Cycle	Total
1959	832	223	83	18	12	1168
1960	868	216	85	21	11	1201
1961	882	209	89	20	10	1210
1962	905	215	94	20	9	1243
1963	884	191	95	18	7	1195
1964	887	191	98	16	7	1199
1965	884	180	99	13	5	1181
1966	876	175	100	11	4	1166
1967	884	172	98	9	4	1167
1968	840	167	105	8	3	1123
1969	856	157	100	7	3	1123
1970	848	151	103	6	3	1111

Source: London Transport Board

Notes 1: For the purpose of the road counts on which these figures are based, Central London is defined as the area bounded by Grosvenor Gardens and Park Lane in the west, Marylebone Road and Euston Road in the north, Shoreditch and Aldgate in the east and the Thames in the south.
2: There are various estimations involved in arriving at these figures, which should be regarded as approximate only.

clear statement that the former objective has been abandoned. It refers to the ringway plans and to the Council's public transport and parking policies and goes on to say:

The task of planning Central London's road system within the new conditions has yet to be more fully tackled, and the developing structure of Central London will largely depend on the secondary network throughout the area enclosed by Ringway 1. Its improvement requires to be studied in conjunction with environmental planning and traffic management, and should promote the segregation of pedestrians and vehicles wherever possible.[48]

But the evidence given by various GLC officials at the Greater London Development Plan Inquiry has made the position rather more definite.

Most people use public transport to travel to and from Central London and to move about once there. For that limited area this is clearly the right answer to the problem of conserving the national heritage areas. There may be some scope for better ordering of vehicular movement for local advantage but there is little possibility for change that is relevant to the essential strategic problems of London. Central London is an extraordinary phenomenon and likely to remain so.[49]

It is not intended that there should be a singificant growth in central London's road traffic and the Council does not propose to provide an entirely new road system in central London or to make substantial additions to the existing network.[50]

This change of view has been so gradual and inexplicit that it is perhaps not surprising that its importance has not been fully realised and its implications have not been thought through. Consequently, policy is in a thoroughly inconsistent state. The plans for Piccadilly Circus and Covent Garden still stand, deprived of their basic justification. Although it is not intended to build more roads in Central London, it is intended to spend £60–90 million on traffic management schemes there.[51] It is admitted that in the long term these schemes will neither increase traffic speeds nor benefit pedestrians; the only effect will be to increase the flow of all kinds of motor traffic indiscriminately.[52] And although parking policy is seen as the key method of control, it is proposed to make a substantial increase in the number of parking spaces provided in the centre.[53]

7. London: The Motorways

Abercrombie's plans and their modification were followed by a host of official and unofficial attempts to devise road plans for London which all came to nothing. The next event which needs to be considered here is the London Traffic Survey, in its later stages called the London Transportation Study, which was started in 1962. The LTS (as it is generally called) was a study of the kind described in Chapter 3. It was not based on any new analysis of the problems, and produced no new ideas for road networks, or at least none that reached the point of being tested. All the networks tested were more or less powerful versions of Abercrombie's ring and radial design.

But although the LTS contained no new analysis, its results gave rise to new arguments in support of the proposed network. At the final stage of forecasting, when traffic for the design year (1981) was assigned to the proposed network, it was found that the network would be full – indeed over-full – and that the journeys which would fill it were of a kind different from those which had pre-occupied transport planners in the past. The journeys were not concentrated on the centre, but took a more orbital direction, round the rings of the proposed network. And the predominant purposes were no longer journeys to work but journeys made for a whole variety of other purposes.

The change of emphasis inherent in these findings was clearly recognised at the time. The following passages come from a report of the Planning and Communications Committee and the Highways and Traffic Committee of the GLC dated 19 March 1965.

Although the work of the London traffic survey is not yet complete, sufficient information is already available to give an indication of the scale and pattern of the future demand for travel. An urgent task facing the Council will be the development of a policy and a plan for meeting such demands. Such a policy may well entail a change of emphasis from the attention given in the last few years to radial routes to one of ring or diverting routes.

Roads – the minimum need
The preliminary results of the survey show that the feature which would be likely to provide of itself the greatest return to the community is a 'box' of major roads around central London, linking the existing radial roads and the projected radial motorways. Such a system would encourage traffic having destinations outside the central area to avoid it and would lessen the duration of journeys within it; more important, it would provide for the very many journeys which are not 'local' but which by their pressure along existing main roads congest suburban town centres – journeys of the type Dagenham to Wandsworth, Greenwich to Hampstead. . . .[1]

Another report of the same two committees of July 1966 both developed the theme of orbital or inter-suburban travel and had more to say about future journey purposes.

But clearly neither the peak-hour journey to work in the centre, nor off-peak journeys to and within the centre, are the dominant feature of London's transport problems of the 1980's. It is the very large increase in movements by road outside the centre for all purposes except the journey to and from work, and in particular the increase of $2\frac{1}{2}$ times in the number of non-work journeys made by car. This increase is more than a reflection in transport terms of increasing business activity – it is also and more significantly a symptom of a profound social change, the coming of an age in which an overwhelming majority of citizens will have access to motor cars as their personal means of transport and the will to use them. The opportunities and benefits which freedom of movement can confer could by then become quite generally available in the community, whole new sections of the population experiencing for the first time a widening of their horizons. When to this is added the forecast general increase of 80% in goods vehicle movements, the problem of accommodating even a substantial part of the total traffic becomes still more formidable.

For while the problems of movement to the centre of London are eased by the considerable legacy of railways which radiate from it like the spokes of a wheel, these movements outside the centre, which are distributed over the whole area and have nothing like the same local concentration as we find in the centre, must continue to go by road. And the existing road system will just not accommodate them, amounting as they do to an increase in this area in journeys by vehicles of all kinds at all times and for all purposes of $2\frac{1}{4}$ times.

Orbital roads – The findings of the survey during the last year support initial appreciations of the vital part which can be played by the 'motorway box'. There is, however, a clear indication of the predominantly inter-suburban nature of the traffic from the fact that less than a fifth of the traffic using this route would be destined for the central area. It offers both direct routes across the suburbs and a good alternative for many journeys which now pass through the streets of inner and central London. . . .

But the survey indicates that there is a far more extensive demand for orbital movement than the 'motorway box' could serve. On the north side of the river the present North Circular Road is well aligned to form a logical complement in the overall pattern to the northern segments of the box, although the traffic levels to be expected foreshadow the need for much greater capacity. Predicted traffic figures confirm long-held views that south of the Thames a counterpart to this road is of great importance. . . .

Outside the denser urban areas there still remains a heavy demand for cross-suburban links. Parts of a third ring located on the outer fringes of the built-up area (based on Abercrombie's D-ring) have been a feature of trunk road planning for many years and it would now appear that completion of this ring would be a valuable addition to the London system. It would link the large outer suburban centres and would enable some traffic originating outside London to avoid the more heavily-used roads of the built up area.

With these orbital roads there will need to be appropriate cross links, some entirely new and others based on the motorways existing or proposed which lead towards the centre.

Meanwhile we have adopted as our planning objective the creation in London of a primary road system of urban motorways with a basic orbital pattern of which three separate rings (the D ring, an improved North Circular Road and a new South Circular Road and the motorway box) will form essential elements; and we have instructed the officers to report on priorities on alignments and on social and economic costs as soon as may be practicable.[2]

Were the forecasts to which these passages refer reliable, interpreted simply as predictions of what traffic would use the proposed roads if they were built? There are very grave doubts about that. The models used to make the predictions contained a number of arbitrary and implausible features; it is doubtful to what extent they had been successful in reproducing the 1962 situation, even after a number of *ad hoc* adjustments. The methods used to predict goods movements were based on a misunderstanding of the trend in the 1950's. The methods used to forecast personal movements implicitly attributed to the working-class car owners of the future the habits, tastes and scattered pattern of relationships of the largely middle-class car owners of 1962. The competitive interaction between cars and public transport was not properly represented, and the public transport policies assumed for the purposes of the test are not necessarily those that will or should be followed. There is little doubt that the roads if built will ultimately fill up; they will generate traffic in the various ways that new roads do. But there are many reasons for doubting whether they will be used in the way that the GLC predictions suggest.[3]

In the present context, however, the question that matters more is whether the predictions, if sound, constitute a justification of the motorway proposals. It is clear from the extracts just quoted that the GLC was still appealing to the principle that 'all traffic demands should be met'; it is of particular interest that according to the July 1966 document the objective to create the proposed motorway system was adopted before the costs had been calculated, notwithstanding the claim in the earlier document that the motorway box (Ringway One) would provide the greatest 'return' to the community. It is also clear that the GLC was still treating the traffic flows predicted as the output of the transportation study as forecasts of demand: the fact that the motorways were predicted to be full, and full with orbital journeys, was taken to be sufficient grounds for saying that there was a demand for orbital journeys. But the reason that the pattern of journeys predicted by the study was an orbital one was simply that the pattern of all the networks tested was orbital. If a radial network had been tested, or a grid network or a network of any shape, it is almost certain that they also would have been predicted to be full. This result therefore provides no valid reason to suppose that orbital roads should be built in preference to any other sort of network; in drawing such a conclusion, the GLC and their consultants were deceived by their own techniques.

There is, however, another line of argument which appears to support the idea that orbital roads should be provided. Two quotations from the official documents will illustrate the reasoning.

Of perhaps equal significance is the pattern of distribution of trips involved. Even in 1962, although there was a strong element of work trips oriented to the Central Area – especially by rail – the preponderance of trips were of a random nature. The change between 1962 and 1981 is expected to comprise a very substantial increase in 'non-work' trips – being of the order of 80% compared with an increase of 10% in work trips. By their nature, these trips will increase the proportion which are made between areas outside the Central Area.[4]

Within Ringway Two the travel demands forecast for the 1980's by the London Traffic Survey indicate large increases in the number of non-work journeys by car outside the Central Area. An orbital route just outside the Central Area can serve the twin requirements of providing for these diffuse inter-suburban movements while relieving the Central Area of through trips.[5]

These passages suggest that because the purpose of the new type of journeys are such that they will not be made from the suburbs to the centre, they will be made from one suburb to another, and presumably between suburbs which are far enough apart to justify a possible diversion to a motorway, rather than the use of the existing streets. But of course there is another possibility, that the journeys might be relatively short and local in character. An examination of the journeys that are expected to increase, purpose by purpose, suggests that some of them, at least, must be fairly local, and none appear to be inherently inter-suburban. This is shown in Table 8.

The LTS, in fact, provided no backing for the idea that there was any particular demand for inter-suburban trips, or that there would be any particular dissatisfaction or frustration if the corresponding facilities were not provided. Neither does commonsense support this idea. A Londoner who has easy access to his own suburb, the immediately adjacent suburbs and the central area is immensely rich in facilities of almost every kind. There are still many problems and disadvantages of living in London, as in any big city, especially for children and for people bringing up children, but inter-suburban motorways would do nothing to alleviate them.

Even if there were much more substance in the idea of inter-suburban non-work travel demand than in fact there is, it would still be debatable how much attention should be paid to satisfying it. The journey to work or to school is still the main problem in London, as it was in Abercrombie's day, and for many people it is still a long, tiring journey made in very unpleasant conditions. But the LTS systematically diverted attention away from this obvious problem. A study of this sort has little to say about the quality of travel facilities; as explained in Chapter 3, such a study is above all a test of capacity. However uncomfortable, costly and time-consuming the journeys made in London in 1962 were, they were in fact performed; by

TABLE 8

Journeys made by Londoners in 1962 compared with forecasts for 1981

Units: Thousand journeys

PURPOSE	DATE		*Increase expressed as percentage of 1962 figure*
	1962	*1981*	
Work	5377	5907	10%
Personal business	1384	2832	105%
Social	812	1415	74%
Shopping – convenience	660	1250	90%
School	550	1225	123%
Other home-based	1210	2047	69%
Non-home-based	1340	2017	51%
Total	11333	16693	47%

Source: LTS Phase 2, Table 13.5.
Note: The journeys relate to an average weekday (twenty-four hours) but do not cover all journeys made on foot or by cycle.

definition the transport system, therefore, had the capacity to accommodate them. By the test of capacity, the only way in which the facilities existing in 1962, if properly maintained, might become inadequate in the future, would be if the number or pattern of journeys were to change significantly. But in London the size of the population and the location of homes and work places were expected to remain stable, broadly speaking, hence the number and pattern of journeys to work would also remain stable, hence the facilities that were adequate to cope with journeys to work in 1962 would remain adequate for future years.

This line of argument is correct, as far as it goes, but of course it does not show that investment to improve transport facilities in London should be devoted to the aim of providing for additional journeys to be made, mostly for social purposes, rather than to the aim of making existing journeys quicker and more agreeable. No attempt was made to supplement the LTS by a survey to reveal which problems Londoners regarded as most pressing, or to what end they would have liked to have seen the investment put.

Another problem which building inter-suburban motorways would leave untouched is the future travel desires of people who will not have the use of cars. According to the GLC's estimates – and there are grounds for supposing that the methods used over-estimate future car ownership – there will still be 45% of house-holds without cars in Inner London in 1981.[6] In addition, there are the various people within car owning households for whom the car is not available: the old, the young, those otherwise incapable of driving, and people who cannot use a car because another member of the household requires it. The LTS again systematically disregarded or obscured the position of those without the use of cars. As pointed out in Chapter 3, the whole method of analysis, being based on households and not on individuals, glossed over the existence of people in car owning households without cars available.

The fact that members of non-car owning households made far fewer journeys than members of car owning households, even after allowing for differences in income, household size, etc, was remarked, but was interpreted rather strangely. The most obvious interpretation would be that people without cars had much the same travel desires as people with cars, but that these desires were frustrated by the shortcomings of the available public transport system. Since so many people would still have been without cars at the design date, and even more in the long interval between the survey date and the design date, the con-clusion that would follow from this interpretation would be that the public transport system should be radically improved. But the interpretation actually given was that it was the owner-ship of a car which in some way created a demand for more travel, and that public transport not only did not, in 1962, but could not, however improved, create or satisfy a similar demand.[7] Such an assumption was also implicit in the predictive methods used, which were based on the idea that the acquisition of a car would immediately increase the number of trips made by a household but that no improvement in public transport facilities whatsoever could generate a single extra trip on the part of those who had no other means of transport available, and that those who did have the choice between a car and public transport would be completely unresponsive to any improvements in the reliability, cost of comfort of the service offered by public transport. There is, however, no justification whatever for these assumptions, which are not only implausible but contrary to experimental and survey data from many places.

By November 1967, the review of costs which the Council had asked for was complete. From this review it emerged that there was no chance of finance being available to complete the whole motorway network as planned by the original LTS design date, 1981. In fact, without some increase in the grant given by the Ministry, less than 70% of the required sum would be available by 1983. The need to provide a primary road network was considered to be so urgent that it was suggested that some of the roads should initially be built less wide than they would ultimately be. This method of two-stage construction was admitted to be expensive and to create other problems, but it would have allowed the benefits from the network to be obtained earlier. The exact date when the initial network would be completed was not specified, but all the work would have been programmed to start by 1983. It was expected that the ultimate network would have been completed by the mid-1990's.[8]

The feasibility of this programme was soon called in question, and in January 1970 a revised programme was put forward. It appeared that the costs had previously been underestimated, and also that more expensive methods of construction than had originally been envisaged would be necessary in order to integrate the roads successfully into the environment. Apart from the costs, the shortage of skilled staff and the time taken to go through the various legal and other procedures involved in road building made it impossible to adhere to the original programme. The same ultimate network was proposed, but the phasing of the construction was radically altered. The idea of building in two stages was abandoned, the order in which the various roads were to be built was altered and the whole programme was much extended in time. Thus it was expected that only the first phase of the construction of the primary network (a phrase which, however, accounted for about three-quarters of the proposed expenditure) would be completed by the mid-1990's. No date for the completion of the entire network was given.[9]

The present plans, as put forward in the GLC's evidence to the Inquiry into the Greater London Development Plan, are very little altered from those of January 1970. The cost of the primary network (the motorways) is estimated at £1,400 million, spread over some thirty years, and investment on the secondary network is planned to be about half that.[10]

Almost any plans can be made feasible by spreading them over a long enough time, and it is probable that the GLC's amended programme does represent a feasible task. But this extension of the programme destroys some of the original arguments for the motorways; in particular, the idea, still appearing very prominently as recently as November 1967, that the provision of a primary network was a matter of very great urgency. The absence of response to the postponement is an indication of how little anybody really believed the argument that London would seize up, or its economy be irretrievably damaged, if roads were not provided. And even if the provision of motorways would in some way solve London's transport and environmental problems, the postponement of the completion date raises in an acute form the question of what will happen in the meantime, an interval of thirty years.

If the investment programme is feasible, the question still remains whether the expenditure of such large sums of money on London's roads is justified. In 1967, it was claimed that the investment would produce a rate of return of 20%. Details of the calculations leading to this figure were not provided, but clearly they were based on the low estimate of costs put forward at that time. Also, the rate of return was calculated on the entire proposed network, so that it took account of the much less expensive and contentious motorways proposed in outer London as well as the more expensive and damaging inner motorways. During the Inquiry into the Greater London Development Plan, which is still proceeding, attempts were made to compare a much smaller road network, which included only the outer motorways, with the complete network, in order to see what rate of return could be attributed to the inner motorways. The GLC admitted to one objector,[11] the London Borough of Camden, that the rate of return was probably no higher than 5·5%, which is far below what is normally required for new roads; since this admission was made several important new criticisms of the GLC's figures have been put forward.[12] In fact, it can be seriously argued that the rate of return on the inner motorways is zero or negative, in other words that the motorways would be so damaging in various ways that they should not be built even if to build them cost nothing.

The case for the proposed motorways is clearly affected by the GLC's current proposals for public transport investment. Public transport had never been the direct responsibility of the LCC or GLC until 1969, when the Transport (London) Act gave the GLC powers of control over London Transport, and also enabled it to offer subsidies to British Railways for certain purposes. For this and other institutional and technical reasons, very little regard had been paid to public transport investment or management problems when the road plans were being drawn up. The Written Statement of the Greater London Development Plan contained few specific suggestions for public transport and no cost estimates. The evidence on transport submitted by the GLC at the Inquiry was still very unspecific with regard to public transport, but it gave an idea of the order of magnitude of the investment envisaged, which amounted to some £1,300 million over a twenty-year period (this figure, however, includes £650 million to be spent by British Railways both in London and elsewhere in the South East).[13] The improvements which an investment on this scale would bring about are clearly likely to have some competitive effect on the use made of private cars, and hence to lessen the case for building roads and the rate of return to be expected from an investment in roads.

Possible investments in road, rail and buses should therefore be examined simultaneously, so that their interactions can be taken into account, however approximately. But the public transport investment now proposed has not led the GLC to reconsider any part of its road plans. The attitude seems to be that the benefits from the two investments can simply be added together, as if public and private transport were completely separate, serving two completely different markets.

Recent arguments for the motorway proposals rely very little, however, on economic or other formal justification, but on more traditional lines of reasoning. A dominant theme is the familiar idea that the demand must be met, and that only motorways on the scale proposed would provide adequate capacity. This argument appears on several occasions in the evidence given by the GLC to the current Inquiry into the Greater London Development Plan; three quotations will illustrate it:

There can be no doubt therefore that the present system of roads is seriously inadequate to deal with the additional demand for vehicle use which flows from the economic and social objectives of the Plan.[14]

It has been stressed that to forecast the demand does not make it inevitable. In framing policies for car use it is important to recognise how the substantial increases that are forecast flow directly from the social and economic objectives of the Plan. The increases also relate to journey purposes, patterns of movement and times of travel that are difficult to control. When these practical facts are combined with the desirability of maintaining and improving personal mobility it is clear that, even though all the forecast demand need not be realised, a substantial proportion will require to be accommodated.[15]

The forecasts are not, however, inevitable; they represent merely the outcome of the assumptions made. What is inevitable is that the demand for movement will increase considerably and, if London is to continue to be acceptable as the capital city and an attractive place to live in, the great proportion of the demand must be met.[16]

The suggestion in the last of these quotations that without the proposed motorways London would become the kind of place that no-one who had the choice would wish to live in has been put much more strongly in other documents; for example, in an article in *Regional Studies*, written by a senior official of the GLC and the managing director of the firm responsible for the London Traffic Survey, and also in an article in *The Times* newspaper by its transport correspondent.

It is sometimes suggested that as London already has a substantial public transport system and that building new roads is too difficult, new investment in transport in London is not justified. Moreover London's population is falling so the money would be better spent where it is rising.

Surely this is a dangerous argument. People move out of big cities for many reasons. One reason is to gain freedom of easy movement. The problem for London is to improve to the utmost the accessibility of home to work, shop, school, and leisure and between businesses, within financial constraints and without destroying its environment. Its future health as an economic unit and as a place which can provide for the needs of a changing population with new tastes may well depend on the provision of a modern, fast transport system vastly superior to that which exists today.[17]

If the GLC try to do the unacceptable – and I am thinking of too much restraint and too little road – people will simply go elsewhere rather than forgo in London that prize of leisure and affluence: mobility. The downward spiral of high costs, poor environment, and falling population will continue.[18]

There is very little substance in these arguments, particularly as arguments in favour of inter-suburban motorways. Nowhere does the Written Statement of the Greater London Development Plan set out economic and social objectives from which the motorway proposals could be said to flow directly. The passage in the Written Statement that seems to come closest to this is as follows – the environmental claims for the motorways made in this passage will be considered below:

But the Council's overriding aim, in collaboration with the Borough Councils, will be to secure a progressive improvement of the environment so that London as a whole becomes a much more attractive place to live in than it is at present; a place which offers better opportunities for its children and young people to develop, which better meets their needs for physical and mental recreation.

The inconvenience of movement in Greater London and the environmental damage which motor traffic is causing are two problems that are inextricably involved with each other. Movement *per se* is dealt with later in this Statement in Section 5 (Transport). But it can be stated here that the Council's policy has the twin objectives of improving public transport in all possible ways and developing the road system to cater efficiently for the enormous amount of road traffic which is essential to the proper functioning of London.

The Council sees no way of avoiding, over a period of years, a massive programme for the improvement of the road system. The main feature of the programme will be the development of a 'primary network', mainly by new construction (i.e. not using existing roads), and comprising three orbital roads (Ringways 1, 2 and 3) and a number of radials. It is important that the dual purpose of the primary system should be recognised; it is partly to extricate vitally important traffic from the crippling constraints imposed by the existing old-fashioned road system and to concentrate this traffic on modern purpose-designed highways; and equally it is for the purpose of improving the environment in the shopping and residential streets from which the traffic has been removed.[19]

The demand forecasts referred to in the passages quoted from the GLC's evidence are derived from those of the London Transportation Study, which as has been seen, are simply projections of the travel patterns of 1962, without any attempt to examine to what extent the 1962 patterns were themselves necessary, desirable, or in accordance with the wishes of the travellers themselves. Throughout the passages quoted there is the implicit idea that personal mobility and the use of the private car are one and the same thing, so that restraint of traffic would necessarily mean restraint of mobiliy. The suggestion that people are fleeing or may flee from London if new roads to provide them with greater mobility are not built is completely without foundation. Booming house prices in Inner London are evidence of the strong desire of relatively wealthy people to live there. In fact, it is one of the causes of the housing problem in London that better off families are buying houses which used to accommodate many more people.

The passage from the Written Statement quoted above seems to express a fear that traffic which is vital to the economy of London will be 'crippled' in its movements unless new roads are built. The same point has been expressed more strongly and explicitly by the British Roads Federation. 'London needs motorways for its economic survival. If the cost of movement increases with congestion, London will become progressively less attractive to industry and commerce and, in effect, price itself out of its own market.'[20] The effect of the motorways on relieving congestion is considered below, but this is in any case a very weak argument. Even according to the GLC's estimates, which predicted an increase in commercial traffic much greater than seems plausible, only a small proportion of the benefit from the inner motorways is attributed to goods vehicles.[21]

The arguments that the motorways are necessary in order to give a hierarchical structure to the road system as a whole, and in order to separate long from short journeys, still figure prominently in recent official documents, for example:

The road system today lacks any form of hierarchial structure to distinguish the functions of different parts. A hierarchy of roads must be developed in London consisting of primary roads, secondary roads, and local roads.[22]

A functional structure or hierarchy of roads is essential to segregate the often conflicting uses to which roads are now put. The most efficacious way of providing the extra capacity required is by means of a system of primary roads built to motorway or near-motorway standards.[23]

The major shortcomings of the road system are . . . most roads have to carry all kinds of traffic, whether for long journeys, short journeys, local access or general distribution.[24]

If Ringway 1 were not built, there could be no segregation of short or long distance traffic on the present roads of inner London.[25]

However, it was argued in Chapters 1 and 3 that the separation of long- and short-distance traffic and the attainment of a hierarchical network structure are both spurious objectives.

A particular argument for the innermost motorway, Ringway One, is that it would remove a large proportion of through traffic from Central London.[26] This might be a desirable objective in itself, but it is hard to see why it is worth achieving if the traffic then has to go on a longer route through densely populated parts of Inner London. There is also some doubt whether, or in what conditions, the traffic would in fact go round rather than through. It is said in the GLC's *Report of Studies* that 'a prime function of Ringway 1 is to remove 25% of the traffic from roads in the ten square miles of the central area given present levels of congestion'.[27] If present levels of congestion were maintained in the central area, through traffic might well prefer the longer but faster Ringway route. But if present levels of congestion were maintained, presumably because other traffic replaced the diverted traffic, it is hard to see what gain there would be to the central area – certainly, there would be no environmental gain. In any case one of the GLC's principal objectives for Central London is to bring about a considerable increase in average traffic speeds, primarily in order to help buses. It is intended that speeds on central roads should never drop below fourteen miles an hour.[28] It is hardly possible to believe that such speeds could be achieved, but if they ever could, through traffic would be attracted back again.

The fact that there is now a national programme of building inter-urban motorways, which tend to focus on London as the largest city, has been used as an argument to build motorways within London. This argument has been expressed most strongly by Professor Buchanan in his study on North East London.

Granted the case for additional road space, it might then be argued that it should be provided in some form other than the construction of a primary network. We think it is worth recalling in this connection that the country has embarked on the development of a national motorway system. . . . The programme appears irreversible for it represents the country's deep commitment to the motor vehicle as the backbone of its transport system. The choice before us is either to stop the national motorways at the edges of the built-up areas and allow the vehicles to make their way into those areas as best they can through the existing street systems, or to accompany the national motorway programme by a matching provision of urban motorways. These two elements – the national or inter-city motorways and the intra-city motorways – seem to us to be complementary. It is difficult to picture one without the other. The *rationale* behind the system lies, of course, in the fact that the vast bulk of the journeys that take place on the inter-urban system have their origins and destinations *inside the urban areas themselves*.[29]

It is not clear in what sense urban road systems, and London's in particular, are supposed to 'match' the inter-urban motorway system. Granted that most of the vehicles on inter-urban motorways are bound for some city, when they get to that city, particularly if it is London, they represent a tiny fraction of the traffic circulating there. Sooner or later, they must come on to the existing street system, since there is no other way for them to reach their final destinations.

The convergence of national motorways on London has also provided a new argument for the idea that there should be a primary road network within London with a ring and radial shape.[30] The suggestion seems to be that unless the incoming radials are linked by rings they will just end abruptly as 'stub-ends', but there seems to be no reason why they should not gradually diminish in scale at intersections as they come into the city until they merge into the street system. There may be some argument along these lines for an outer ring, but not for inner rings.

The other arguments for the proposed pattern of motorways are familiar. The main evidence on transport presented by the GLC at the current inquiry follows Tripp and Abercrombie in the idea that a sensible approach to network design is first to find the right broad geometrical shape, and then to fit the shape to the city, making appropriate adjustments where necessary. The advantages of the ring and radial pattern were described as follows:

The ring and radial pattern has been found in many urban areas to provide a more even distribution of capacity against demand. This pattern tends to provide more capacity in the inner areas of the region and less for each unit of area as distance from the centre increases. Since in most urban areas densities and hence traffic movements decline as one moves further from the centre the road capacity is more likely to match the demand.[31]

This argument is clearer and more convincing than Abercrombie's, but it rests on the same idea that there would be no difference in the nature of the transport system as between the centre of the city and its inner and outer areas; hence, the greater density of trips in the centre would require a greater density of traffic and so of roads. But if the journeys made in the centre are mostly to be made by train, bus or on foot, rather than by car, the argument no longer applies. For the same reason, the American studies to which the GLC evidence also appealed to support the idea of the ring and radial pattern are also irrelevant.

Another rather general argument given in the official documents for the orbital motorways is that without them the existing mainly radial pattern of roads is defective in that 'it limits diversity of movement'.[32] But diversity of movement, i.e. the ability to move freely in any direction, is not important. The only diversity that matters is the diversity of choice and opportunity to satisfy journey purposes, which is something the

Londoner already has, even though his movements may be somewhat restricted to certain corridors. In any case, this argument hardly applies to car users, for whom movement in non-radial directions is already perfectly possible. It is the public transport user who is likely to find even quite short non-radial movements very difficult to make. If there is any force in the argument, it indicates a need for new non-radial public transport services.

The argument for the particular pattern of motorways that appears most frequently in the GLC's evidence on transport submitted to the Inquiry into the Greater London Development Plan is that it is necessary in order to meet the new spatial pattern of demand. There is no bold statement of the type found in the March 1965 or July 1966 documents that the new demand will be 'predominantly inter-suburban', journeys of the type 'Dagenham to Wandsworth, Greenwich to Hampstead'. The idea that orbital facilities are what is required is introduced very gradually, almost imperceptibly. It is first claimed that the new journeys will be of a dispersed pattern rather than being concentrated.

The increase in social and economic activity is reflected in both the number and purpose of journeys and in an increasing dispersal of the physical pattern of movement. In the past most personal journeys started or ended at home and the other end has been at a dense focus of activity. With changing social patterns this is changing to a situation where a much smaller proportion of the personal movements has an origin or destination in high density centres.[33]

This statement is supported by numerical forecasts which suggest that there will be an increase of less than one million journeys to Central London and other major town centres in the London area between 1962 and 1991, but an increase of well over four million in journeys to all other destinations. After further argument it is concluded:

Increased personal wealth and widespread car ownership are causing Londoners to participate more and more in social activities, such as shopping, entertainment, education, and recreation, all of which entail the use of transport. Demand for movement for these purposes shows the greatest growth rate, almost doubling between 1962 and 1991. These journeys have widely dispersed origins and destinations and it is this demand which will put the greatest strain on the transport system.*[34]

In the following chapter of the evidence it is argued:

London cannot stand still and the need for some road investment is probably accepted by most people. The real options arise in deciding how much should be spent and whether the money should be used to build motorways or to improve the existing system. . . . To improve the existing roads would intensify the conflict between traffic and environment and possibly remove large parts of highly valued property in the process. Furthermore improvement of the existing system alone implies increased radial access as there are few non-radial roads. We have shown in Chapter Two the large increase in the demand for journeys *dispersed over London* [my italics].[35]

Hence it is concluded that one of the strategic objectives of the primary road system should be 'to strengthen the general structure of the road system particularly by the provision of orbital routes', and it is later stated that among the considerations that had to be observed in designing the primary road system was that 'orbital communication is required to satisfy the biggest forecast increases in demand'.[36]

The whole of this line of argument contains the fallacy that has already been pointed out in discussing the conclusions drawn from the LTS: it ignores the possibility that the new social travel demand might be predominantly local. Certainly, new social journeys are likely to be dispersed in the sense that journeys made in one part of London will be far removed from journeys made in other parts. But they need not be dispersed in the sense that is required to justify the provision of orbital motorways – namely that the destinations of particular journeys will be a long way from their origins. Where the LTS forecasts suggest that the new social journeys will be dispersed in the required sense, that is, that they will be long orbital journeys, this, as has been seen, is simply a consequence of the fact that all the road

* Even accepting the forecasts on which this passage is based, the conclusion that the social demands referred to will 'put the greatest strain on the transport system' hardly seems to follow; commuter journeys made by public transport are still likely to be the most troublesome. What would follow from the forecasts is that the new social demands would place the greatest strain on the *capacity* of the *road* system.

networks tested in the LTS were of a pattern which encouraged that kind of journey.

Two matters which have aroused increasing public concern in recent years are the environment and the declining service and rising price of the bus system. Recent arguments for the motorways have placed great emphasis on the ways in which they can help in these two respects. As was shown earlier, environmental considerations weighed very strongly with Abercrombie when he put forward his road proposals; but very little attention was paid to them in the LTS. They have recently been stressed in a way which almost suggests that the primary purpose of building the proposed ringways is to improve the environment. The passage from the Written Statement of the Greater London Development Plan which refers to the improvement of the environment as the GLC's 'overriding aim' has already been quoted, and this is given as one of the two main purposes of the primary network, although it is later stressed that the primary network 'cannot be relied on by itself to bring about the full range of environmental improvements'. It has been seen that Abercrombie said almost nothing about buses in any context, and certainly not as a justification for building motorways, nor did any such idea appear in the early stages of the LTS. The earliest official mention of it seems to have been in the GLC's minutes of July 1966, which have already been cited, where it was said:

It must be emphasized that an improved primary road system will also be of great help in maintaining an improved popular service of buses. New services can use the new roads and other services will be able to make much better use of the existing road network.

In the last two or three years, the GLC has increasingly adopted a position in favour of public transport, whether in response to public pressure or because the limitations of a strategy intended to satisfy demands for private transport have made themselves increasingly apparent. In line with this development, the argument that the motorways will help buses has been increasingly stressed. In one recent popular policy document it is said that:

The general improvement in the flow of traffic which will follow the progressive realisation of the GLC's plans for the construction of the primary road network and the improvement of the secondary road network will provide further help to the smooth running of the bus services.[37]

In the manifesto of the Conservative Party for the GLC elections of 1970 it was said:

One of our biggest spending programmes is designed to provide an adequate road system for the bus, for the commercial vehicle and for the reasonable use of the car, by building the Ringways in London. We are committed to pressing on with this task so that we can meet the traffic requirements of the next decades and remove the heavier, long-distance traffic from local roads and town centres.[38]

The direct effect on the environment of building motorways in London, particularly in Inner London, is clearly adverse. The inner motorways and linking radials amount to more than sixty miles of road, not counting sections which are already built or which would consist of existing main roads up-graded. Almost all will be six-lane or eight-lane roads; many sections will be elevated. Whatever care is taken to find the best alignments, the insertion of such roads into a densely built city must be extremely damaging. The motorways will also provide no direct help to the ordinary London buses, which cater for short journeys and require frequent stops. It is unlikely that such buses will ever travel on the motorways, which, indeed, represent a new source of competition for buses. If the motorways are to bring any net improvement to the environment or to the bus system, there must be very substantial indirect benefits to compensate for the adverse direct effects, and this, apparently, is the GLC's hope.

Widespread congestion exacts an unacceptable toll in terms of cost, unreliability and accidents; it particularly affects bus services; it also encourages excessive amounts of traffic to permeate minor roads bringing noise, danger and ugliness where there should be relative quiet. The Development Plan seeks to reduce this debilitating congestion. Road building and improvement, traffic restraint and the improvement of public transport are each necessary to this objective.

The major step in this strategy is the building of a new primary network of roads to a high standard. For with so much depending on relief from congestion, a radical improvement in capacity is needed if the balance in the total environment is to be achievable.[39]

The clear suggestion of this passage, and the other passages from official documents cited, is that when the primary network is built conditions on the rest of the network will be very much better than they are now. This reasoning is, however, quite wrong, and should never have survived the LTS. The initial traffic assignments of the LTS showed overloadings on the secondary network as well as on the primary network.[40] By definition, such overloadings cannot come about, but this finding indicates that the secondary roads will be under at least as much pressure when the primary roads are built as they are today, as indeed is indicated in another official document, the *Report of Studies*: 'Although construction of the primary network is aimed to relieve the secondary network of much through traffic the sheer overall growth in traffic is likely to result in vehicle flows on the secondary roads being as heavy or heavier than today.'[41] Since the secondary roads are also bus routes, there are no grounds for supposing that the present difficulties which congestion imposes on bus operation will be lessened when motorways are built. In so far as almost all local shopping and other centres in London are traversed by secondary roads, this finding also shows that no environmental improvement can be expected in such centres.

There remain, however, several thousand miles of local roads on which the environment might be improved; such improvements would be very important since these are the streets in which most Londoners live and where many, particularly the very young and the old, spend almost all their time. The main part of the LTS threw little light on this problem, since only the main road network was examined. But a detailed study subsequently undertaken in one locality to throw light on these problems found that although the zone in question was already so affected by traffic that its environmental quality was unsatisfactory in almost all respects, and although it happened to be an area from which the new roads would remove almost all the through traffic, the increase in locally generated traffic would be such that the overall situation after the new roads were built would be little changed.[42] The *Report of Studies* suggests a similar general conclusion. 'Local movement is expected to increase significantly and road space will be under pressure from the competing demands of movement, delivery and parking.'[43]

The GLC's objectives, as formulated, will therefore not be achieved by building motorways, but it may be said that, nevertheless, a comparison between the present state of affairs and the state of affairs when the motorways are built is not really relevant. The only comparison that matters is between what the future would be like if the motorways were built and what it would be like if they were not. The GLC clearly expects the 'with motorways' situation to be much better than the 'without motorways' situation, for reasons which are made clear in the following passage from an official GLC document:

The relatively high standards of capacity, safety and speed of primary roads will encourage the transfer of inter-suburban and inter-regional journeys to them from the existing roads. It is not possible to estimate just what proportion of the traffic would transfer to the primary system. Certainly it will be substantial; conceivably the primary system when complete will take as much as half the total vehicle mileage.

It is this shift of a substantial portion of the total movement from the existing road system which will provide the opportunities for higher local standards both of environment and service. . . .[44]

This argument seems to be appealing to a very simple theory of traffic behaviour, that the number and pattern of vehicle journeys to be accommodated on the road system is determined independently of the particular roads that are provided: the only effect of building new roads is to provide new routes for the traffic and hence – if the new routes are more attractive – to divert it from its previous routes. Of course, new roads do have some such diversionary effect. The proposed motorways would certainly cater for some journeys which would otherwise be made by the existing road system, although any vehicle diverted on to the motorways would still have to travel along existing roads at each end of the journey.

But there are also the generating effects of the motorways to be considered. They are intended to cope with more vehicle journeys than are made now or could be made on the existing road system. Each extra journey generated by the motorways also imposes an extra load on the secondary roads which provide access to the motorways and on the local roads on which the vast majority of journeys begin and end. This effect is likely to swamp the diversionary effect, and even if it did not, any relief brought by the motorways to the rest of the system would soon be absorbed by more local traffic generation. Hence it is quite probable that there will actually be more traffic on the rest of the road system if the motorways are built than if they are not.

The GLC's own forecasts are difficult to reconcile with the idea that there will be any less, since, as has been seen, the assignments showed overloads both on motorways and on other roads.

The conclusion that should be drawn from the LTS and subsequent studies is almost exactly the opposite of the conclusion that the GLC and like-minded people have drawn. They have argued that the problems of traffic and the environment that London faces have been shown to be so colossal that only a corresponding effort in road building can solve them. Even the GLC's proposals, which would extend over thirty years, cost £2,000 million and involve the destruction of 20,000 homes are barely sufficient. The conclusion that should be drawn is that this is a bankrupt strategy for London. It will not prevent, and may even help to bring about, a further deterioration of the transport system and the environment.

8. London: The Alternative Strategy

Clearly it is beyond the scope of this book to set out detailed transport proposals for the whole of London, but it is possible to outline a strategy for the central and inner areas.

The strategy for the centre is relatively easy to formulate, because the constraints imposed by existing development and by the role played by the centre are so extreme that they leave very little room for manœuvre. All the considerations – environment, transport, cost, conservation – point in the same direction. Once one is freed from the false ideas and misconceptions of the traditional approach, as described in Chapter 1, the right direction for policy becomes obvious, and the elaborate studies of recent years have added little except further confirmation.

The situation in Inner London is less constrained and the strategy is therefore less obvious and the studies that have been done (or something similar) were necessary to clarify the possibilities. With their aid, it is possible to formulate a strategy for Inner London. In Outer London, the situation is much more open. Perhaps indeed it makes no sense to talk about Outer London as an entity. The problems of the various townships that make it up should perhaps be studied one by one. That task will not be attempted here.

Central London

At present a vast number of personal journeys are made to, from and within the centre of London. If the centre is to continue to play anything like the same role in the life of the city and the nation as it plays today, a comparable number of movements must continue to be made there. It is not possible, even if it were desirable, for more than a small proportion of those journeys to be made by individual motor vehicles: cars, taxis and motorcycles.

The aim should therefore be to improve the alternative means of transport as far as possible, both for the sake of the majority of travellers who must perforce always make use of them, and to provide the minority who now use cars and taxis with a satisfactory and, if possible, a preferable alternative. There will always be a residue of journeys which even a greatly improved shared transport system could not cope with, so some place must be found for cars or taxis. But the aim should be to reduce to the minimum the reliance placed on such means, and never to allow the volume of cars and taxis to increase to the point that they impair the efficient and reliable working of the alternatives, as they do now.

The alternatives certainly include the pedestrian system as well as buses and trains. It is useful to think of Central London as a large building, which one would not enter with a car. Movement within it would be either on foot or by the specialised means of transport provided – both methods should be agreeable and convenient.

There are important differences from a large building, however. One is that many people would find it convenient to be able to move about in Central London by bicycle. The problem is that bicycles do not mix very well either with pedestrians or with motor vehicles and it is hard to know, since this problem has never been studied, what would be involved in making special provision for them. It would seem that many features of a policy which was designed to aid pedestrians and bus users would incidentally help cyclists too, but special studies would be required to investigate whether it would be possible to help them more directly than that.

Another much more important difference from a building is that one can use the building's specialised transport facilities (usually lifts and escalators) only after one has entered it. But the buses, trains and taxis which carry travellers about within Central London also extend outside the centre and are the main means of bringing people there. It is desirable that travellers to the centre of London should get onto the specialised system as far out as their journeys permit, both in order to avoid massive interchange problems at the boundary of the central area and to eliminate unnecessary vehicle mileage on the streets of Inner London.

In general, one would expect travellers from Inner London to the centre, other than the minority who walk or cycle all the way, to make their whole journey by train or bus. But for many travellers from Outer London or beyond that would not be possible; they will have to start their journey by car. They should be encouraged to change to bus or, much more commonly, to train as far out as possible by the provision of car parks at stations in the outer areas, by fast train services and if necessary by the railways' pricing policy, so that fares would not automatically increase with the length of the journey. At the same time, they should be prevented from bringing their cars into the inner areas or right into the centre by parking restrictions or other deterrents.

This policy may require some investment in extra rail capacity in particular parts of London's rail system, together with investment in the quality and comfort of the system. It will certainly require investment in railway and Underground stations, including stations in the centre, to make them more pleasant places, to replace the antiquated lifts and entrances in many stations and to facilitate the interchange between trains, buses and taxis.

The shared specialised system for movement within the centre consists of the Underground and the bus system. The immediate and vital problem is to improve the buses. Many journeys within the centre are relatively short, and journeys are made in all directions, which means that they do not lend themselves to trains, which are better suited to cope with comparatively long journeys along a corridor. An improvement in the bus service would attract some people who now travel by Underground, thereby making conditions easier for the remainder. It requires only management measures to bring about a substantial improvement in bus services; such measures do not cost very much and can be reversed. Any considerable improvement in the Underground services, however, is likely to require a substantial investment in new facilities. It is a sound rule to see what can be achieved by inexpensive and reversible measures before committing oneself to expensive and irreversible measures, especially as studies on even London Transport's most favoured schemes indicate a very low rate of return on new Underground lines.

There may well be room for a revision of the bus routes in

London, many of which have remained substantially unchanged since the time of the bus companies which preceded London Transport. But the immediate need is to increase the speed, frequency and reliability of the existing services. This requires easier conditions for buses on the streets. Easier conditions could be brought about either by a general reduction of traffic volumes, so that buses, in common with the rest of the traffic, would enjoy a relief from congestion; or by reserving lanes or streets for the use of buses (perhaps in common with a limited number of other vehicles, such as delivery vehicles) so that buses would be protected from the congestion elsewhere on the network. Such priority measures would, however, imply some reduction in the total volume of traffic, since it would reduce the amount of road space available for other kinds of vehicle.

Pedestrians require to be able to move freely and easily over the whole of Central London. Of course, very few individual pedestrians are likely to want to walk right across the central area, but some individual journeys will be fairly long and different journeys overlap with each other in a way that makes it impossible to cut up the central area into a lot of small self-contained environmental areas. To provide adequate facilities for pedestrians will entail some widening of pavements and perhaps some conversion of existing streets and squares for exclusive pedestrian use. It will also mean making it easier for pedestrians to cross traffic routes. Crossings should normally be at ground level, since any change of level is inconvenient and disruptive for people on foot. For safety, ground level crossings should be controlled by lights, and if the lights are to favour pedestrians for a reasonable proportion of the time, the effective capacity of the road system to carry motor vehicles is likely to be reduced, although perhaps not by more than is anyway required to facilitate the workings of the bus system. On very busy traffic routes some pedestrian crossings will have to be provided above or below ground level; such crossings should make use of escalators or moving travelators wherever possible.

What methods of control should be used to ensure that the traffic on the central streets does not build up to a level which interferes with bus and pedestrian movement? This question is closely bound up with the question of how provision is to be made for the residue of journeys that will still require individual means of transport even when the shared system has been improved to the uttermost. There must always be taxis to provide for a large part of such journeys, since many travellers will not have cars available to them. It is a tempting idea that taxis should be the only means of individual motorised transport within the central area; in other words that no cars should be allowed in the centre.

The first great advantage of such a system would be its simplicity. The rule would be quite clear cut, and since taxis and cars are easily distinguished from each other it would be easy to administer. The total number of taxi journeys could be kept in check, to the limit set by the good operation of the bus system and the maintenance of satisfactory pedestrian conditions, by financial measures, making use of the taxi meters which already exist. To rely solely on taxis would immediately eliminate the parking problem. It would also have great environmental advantages. Traffic volumes could be kept down in streets where to do so would be particularly desirable, simply by proscribing their use to taxis unless their journeys began or ended in them or in one of the adjacent streets. Such a system would be much harder to work for cars, since there would be more vehicles involved, car drivers could not be expected to know the street system and the rules well enough, and it would be difficult to devise or enforce suitable penalties. London taxis are already specially designed vehicles, subject to very strict standards of driving and maintenance. It would be possible to add further design specifications particularly for environmental reasons. For example, London's taxi service might become the first major application of the electric car.

There are, however, a number of problems about excluding cars altogether from the central area. The first is the problem of the car owners who live in the centre. It could be argued that it should be an accepted fact of living in the centre of London, or at least in that part of London between the West End and the City, that it is not possible to keep a car there. The great attractions of the area, which would be considerably enhanced by the removal of cars, are sufficient compensation. The second problem is that of through traffic. The improvements to public transport required to facilitate access to the centre will also allow some journeys from one side of the centre to the other now made by car to be made by public transport. However, there will still be need for some car journeys, and journeys of goods vehicles too, to be made from one side of the centre to the other. One solution might be to define certain routes which could be used by all classes of through traffic, as well as by the buses, taxis and goods vehicles serving the centre. One appropriate east–west route for the purpose would be the Embankment; another would be the line of Westway, Marylebone Road and Euston Road. North–south routes connecting with the river crossings would also be required.

A third objection to the idea of banning all cars in the centre would be that it would make some journeys needlessly more difficult and expensive. For example, a journey made from Inner London to Central London in the evening to visit friends, or at weekends to one of the central museums or parks, is most easily made by car, and there may be no harm in allowing it to be made by car. One answer to this could be to have rules in the evenings or at weekends different from those which apply during the working week. Or it might be said that some greater expense or inconvenience for a limited number of journeys is part of the cost of adopting a system which would be generally superior.

There are certain advantages in allowing cars as well as taxis in the centre, but it poses new problems of control. Parking control is a powerful and perhaps adequate method of limiting the total number of journeys made by car to the centre, but the problem is that through traffic entirely escapes it. Hence any relief which parking control might bring to the central streets would probably soon be taken up by an increase in through traffic. Parking control therefore needs to be supplemented by some method of control which would bear on through traffic. One such control might be a barrier along the line of the river and another running through the centre in a north–south direction which could not be crossed by private cars during the working day. Another possibility would be some form of road pricing. Perhaps the most convenient and easily introduced road pricing scheme would be a payment to cross a cordon surrounding Central London. An advantage of this would be that it would bear on cars destined for Central London as well as on through traffic, and hence might ultimately replace parking control.*

* A method of charging by the day for entry to Central London has been described by J. M. Thomson.[1]

It is always the exceptions and the odd cases that cause the most trouble in designing any kind of management or control system, and clearly a great deal of thought would have to be given to the exact way the Central London system would work. The ideas just described are suggestions only. Perhaps the most striking thing about the general system proposed is how closely it resembles the system that already exists. Central London already has its own specialised transport system and the great majority of journeys either make use of it or are made on foot. But the conditions for the great majority of journeys are seriously affected by the minority of journeys made by other means. The problem is not to invest more in order to increase the number of vehicles that can be accommodated, nor to change the character of the transport system in any way, but simply to purify the system and allow it to work properly. This is a problem of management and nothing else.

Inner London

The position is not so clear cut in Inner London as in Central London, but it is possible to say with some certainty what the general strategy should be, although it is not possible to say exactly what management measures or investment are required or to what extent various desirable aims might in practice conflict with each other. The main considerations to be taken into account are as follows:

1. *The majority of people in Inner London are without the use of cars. This situation is bound to continue for a number of years and may always be so.*

The 1966 Census showed that less than 30% of the households in Inner London owned cars. The GLC forecasts suggest that by 1981 this will rise to 55%. By the time allowance has been made for people in car owning households who are unable to drive or cannot use the car because another member of the household requires it, it is clear that most people at any one time will not have cars available. Moreover, there are very grave doubts about the forecast. It is difficult to reconcile the methods used with the fact that between 1962 and 1966 the proportion of households in Inner London with cars hardly rose at all, and in many boroughs fell. The forecasts took no account of the fact that it is already difficult to find space to keep a car in many parts of Inner London, and that as ownership increases this problem will become more severe and more widespread. It also assumes that the desire to own a car is quite unaffected by the proximity of schools, shops and other facilities and by the availability and quality of alternative means of transport, matters which are largely determined by public policy.[2]

2. If car ownership were to increase to the predicted levels, it would not be possible to provide the average car owner of the future with the same level of service as that enjoyed by the average car owner today.

This conclusion is strongly suggested by the GLC's own studies; it will be remembered that when the predicted 1981 traffic flows were assigned to the proposed road system it was found that the roads were of insufficient capacity to accommodate them. This showed that the tacit assumption of the prediction was not justified. The proposed road system would *not* provide future car owners, in their greatly increased numbers, with the same level of service as the existing road system offered existing car owners. There are two reasons why this test is not completely conclusive. One is that all the road networks tested were of a kind that encouraged long distance inter-suburban journeys: it is conceivable that tests on networks of different patterns might have showed a different result. The other is that the tests took no account of the relief which would have been brought to roads in Inner London by the adoption of the kind of policy for access to Central London that has been described above. On the other hand, the road networks tested and now proposed represent an extreme policy of road building in that it presupposes that very great sums of money will be available and that environmental and other considerations can be largely set aside.

3. Even the present number of cars on the roads of Inner London is greater than it should be in relation to the requirements of buses, pedestrians and the environment.

This is a matter of observation and experience which is not seriously disputed by anybody.

4. It is more important to improve conditions for the essential journeys to work and school than to provide for an increase in social trips.

It would be desirable to conduct some social surveys to throw light on the priorities which Inner London residents themselves hold. In the absence of that kind of data, this is a judgement based on commonsense and on observation of what shortcomings of London's transport system seem to arouse most criticism from the general public.

These considerations suggest that the main aim of transport policy in Inner London should be to ensure that no resident of Inner London should have to rely on the regular use of a private car for ordinary purposes of movement within Central and Inner London. Hence, the role of the car should be to supplement other means of transport for exceptional journey requirements for which they are inadequate, and to provide for any other journeys which the traveller finds more convenient to make by car and which he can make by car without unduly interfering with the environment or with other travellers.

These principles are not completely specific. It is not clear exactly where the line between ordinary and exceptional travel requirements should be drawn. Doctors and travelling salesmen will probably always require their own cars, but many other journeys which now more or less require to be made by car could be made by an improved public transport system. Cars will be needed not only to travel out of London altogether (and for this purpose many people may find it more convenient to hire cars than to own them) but also to make many inter-suburban journeys, particularly outside normal working hours. The importance of such journeys has been grossly exaggerated in the GLC's forecasts; they are not very important now and no policy should be adopted which would encourage them to increase. But some people do have a scattered pattern of social relationships which would be very hard to provide for except by car. Cars are also very convenient, especially for housewives with young children, for a host of local trips. They cannot be said to be necessary for these purposes, since most people do and must get on without them. But although the car must never become a necessity for such purposes, it does confer an extra convenience, and there is no reason why car owners should not enjoy this extra convenience so long as they do not thereby interfere with other people.

Somewhat unspecific though they are, these principles do indicate very clearly the direction policy should take.

For land use, the policy should be to ensure that as many social, recreational and shopping purposes as possible can be satisfied either in individual localities or in Central London, where the facilities can be reached by travellers without cars. This implies a departure from the present GLC policy of attempting to build up selected 'strategic' centres for shopping and other purposes in Inner London while allowing other smaller centres to decline.[3]

The first implication for public transport is that access to the centre should be maintained and facilitated; this has already been discussed. For local transport, buses are all important. More frequent and reliable services on existing routes should be provided and there is also probably a need for new routes, some of them of a non-radial kind. To maintain the service to the required standard bus lanes and other priority measures will be required, and it may well also be necessary to subsidise the services. Such subsidies should not be thought of simply as a present to the users of the service, but as a means of avoiding the environmental damage and the expenditure on road construction that would be required if people were forced to rely on the use of cars.

Pedestrian movement is extremely important for many kinds of journey in Inner London and there is every reason, economic and environmental, to encourage pedestrians. That is one very good reason for trying to build up the services offered in each locality. Walking should be made safe and easy both to and within local centres, as well as in residential neighbourhoods. Cycling is also important, particularly for children and young people. Parents should not have to become chauffeurs for their children simply because it is unsafe for children to bicycle, and children should not have to be dependent on their parents.

Apart from bus priority measures, the main management measures likely to be required are parking controls. The need for parking controls to deter travellers from Outer London or beyond destined for the centre from driving as far as possible, rather than changing at the outskirts, has already been mentioned. Such travellers usually tend to park on residential streets near to convenient bus stops or Underground stations, and residents' parking schemes are probably the best way of dealing with that. In addition, parking control at local centres will be required both to prevent the number of journeys made to them by car from rising to a number which makes access by other means difficult and to ensure that as far as possible the spaces are used by those who require to use them. This requires control of the total number of spaces provided, of the rules for the times that parkers may come and go and the length of time they may stay, and of the charges made.

Traffic management measures to limit speeds and flows in streets where high speeds and flows are particularly undesirable will be necessary. However, in Inner as in Central London, there are few streets where environmental considerations can be disregarded, and there are therefore strict limits to the gains that can be expected simply from re-routing traffic. Such measures are therefore likely to be most beneficial where it is possible simultaneously to reduce the total volume of traffic.

What part should the provision of new roads play in all this? There will certainly be a need for some new roads, since otherwise it is doubtful whether the various measures described above could be implemented. In particular, to create attractive local centres which are convenient for pedestrians may not be possible without considerably rebuilding them and their road systems. Whether there is any case for building new roads simply in order to increase the total capacity of the road system is more difficult to say, particularly before the effect of all the other measures that have been described has been seen. But if some addition to total road capacity does prove to be necessary, it is certainly not the major feature of a rational transport strategy for Inner London.

9. Managing the Roads

The need for managing the road system has been argued and illustrated in earlier chapters. In the short term, the aim is to redress the bias inherent in the way that the transport system works when there is no intervention: in other words, to improve environmental conditions and to make travel easier and more attractive for actual or would-be pedestrians, cyclists and bus passengers. In the longer term, there is a further object of saving public money, since if roads are not well managed, it is likely that more roads will be required to achieve certain transport aims than if they are. In this field, as in others, good management can reduce the need for investment.

There is a great variety of actual and possible management measures. The purpose of this chapter is to list the main types and to discuss the advantages and limitations of each. No attempt will be made to say which is 'best'. In the abstract, this is a meaningless question. We are not limited to one measure, and such measures are more powerful when used in combination than singly. The problem is to find the right mix of measures, and what that is must depend on the circumstances of a particular town. It would be beyond the scope of this book to illustrate in detail what can or should be done in particular towns, but readers who are interested in finding out more about schemes that have been put into operation in Britain, other European countries and elsewhere will find detailed descriptions in other sources.*

The line between what should count as management and what should count as investment is not always very clear. Some of the

* See, for example, *Can We Kill the Car?* by Terence Bendixson, Temple Smith, 1972. A description of some existing and proposed bus priority schemes in Britain is given in *Working Group on Bus Demonstration Projects Report to the Minister*, Ministry of Transport, 1970. Descriptions of some bus lanes in operation in Europe and the United States are given in *Revue de l'UITP* (International Union of Public Transport), Volume XVIII, No 3, 1969.

measures described below involve physical alterations to the road system, which might be thought of as going rather beyond mere management. But the measures are all relatively cheap, and could, if necessary, be reversed fairly easily; these are the vital things that distinguish them from building new roads, or multi-storey car parks, or other kinds of transport infrastructure. Similarly, subsidies to public transport may not be immediately thought of as a management measure. But they too are reversible measures which are likely to be at least relatively cheap and which may be a partial substitute for building roads.

Control can be exercised over either the moving vehicle or the parked vehicle. A simple method of exercising some control over the moving vehicle is to change the surface of the street so as to force vehicles to slow down, to walking pace if necessary. This has been done in Norwich, by means of a cobbled hump and depression in the road. Such alterations are a convenient method of controlling the behaviour of vehicles and perhaps indirectly limiting their number when it would not be desirable to exclude them altogether. Similarly, at road junctions pavements can be made continuous across the road and can be given sloping rather than vertical sides so as to make it possible for motor vehicles to cross them. This would not only slow motor vehicles down; it would be a useful means of reversing the rule that at intersections it is always the road vehicle and not the pedestrian that has priority.*

Physical alterations to the road surface lead on naturally to alterations in the road network. Roads can be converted into culs-de-sac or in other ways sealed off (in Sir Alker Tripp's phrase) from the road network. As an extreme case, they can be removed completely from the road network, as happens when a street is converted for exclusive pedestrian use. Entrances to streets can be narrowed to make it physically impossible for lorries and other large vehicles to enter them. Bollards or posts can be put at the entrances to streets thereby limiting access to pedestrians and cyclists. Certain people, for example local residents and shop-keepers, can be provided with keys which enable the posts to be lowered to the ground so as to permit the passage of motor vehicles.

* This idea was suggested to me by Dr Mayer Hillman; as far as I know it has never been put into practice.

In this way physical measures of control, which in the first instance discriminate only according to the physical characteristics of vehicles, can be extended so as to discriminate between different users of motor vehicles too. Physical barriers need not be permanent: posts or other barriers can be in place at certain times of day and not at others. In this way a street which forms part of the road network during the working day can be converted into a pedestrian promenade in the evenings and at weekends. A precedent for this is the recent closure of The Mall, in London, on Sundays, to the immense enrichment of the users of the neighbouring parks.

The same effects that can be achieved by physical means can, in principle at least, be achieved by legal* means, although the problems of enforcement might sometimes be formidable. For example, it is possible to impose very low speed limits on certain streets, but it would seem desirable at least to supplement a legal speed limit by some physical measure that would help enforcement. But legal measures can also be used to make discriminations which it might not be practicable to make by physical means only. One example is the exclusion of certain types of vehicle from whole areas rather than from individual streets. There are long-standing precedents for this, in the rules forbidding vehicles over a certain weight from entering town centres which are in force in many towns, or in the ban on commercial vehicles in London's parks. It was mentioned in Chapter 3 that to specify bus routes is in effect a legal means of control, in so far as buses are prohibited from entering streets not on their routes; similarly, taxis can be forbidden from entering certain streets except when their journeys start or finish there.

Certain parts of the road can be reserved for particular types of vehicle, for example lanes for buses, or for buses and taxis only; again it may be a help to enforcement to have low kerbs which would prevent other vehicles from crossing into the reserved

* The term 'legal' is slightly misleading in that it may suggest that the other forms of control described in this chapter do not depend on the law. Control by traffic lights, road pricing and even some of the physical measures described would only work if there were a legal obligation to comply with them. But the classifications used in this chapter are not intended to be very precise; they are only a convenient method of grouping the various possible methods of control for the purpose of discussion.

sections, rather than relying simply on paint or other road markings. If it is desired to deny access to all vehicles to a street at a certain point, it is probably best to do so by a physical closure. But physical closures are not suitable if the aim is only to prevent certain turning movements, as right turns are commonly banned in traffic management schemes. It is possible to give priority to buses, or other type of vehicle, by exempting them from such turning restrictions. Vehicles can also be identified, for the purpose of making legal discriminations, in some other way than by their type. For example, access to certain streets or areas could be allowed only to vehicles whose drivers held a pass, or to vehicles which were themselves identified in some special way. The passes or other means of identification could be granted on a permanent basis to certain classes of people, local residents or property owners, doctors with practices in the neighbourhood, or on a temporary basis, perhaps as tickets which could be bought. These rules, like physical restrictions, could operate at certain times only.

Traffic lights are a long-established means of control. Originally thought of simply as a means of providing for the safe and orderly movement of road traffic and pedestrians at individual intersections, they have been increasingly used for a wide variety of control purposes. The speed at which traffic moves on a main road between intersections can be controlled by phasing the lights in such a way that there is no point in driving fast. This is a very useful aid in maintaining road safety.

Traffic lights are now being used as a means of increasing the capacity of the road system either along a particular road or over a whole area. They can also be used to restrict the flow along a given road so that certain speeds can always be maintained; when the number of vehicles on a given section of the road has reached the limit compatible with the desired speed, traffic lights prevent any more vehicles from coming on. This could be a very valuable means of maintaining the reliability as well as the average speed of a bus service, especially when combined with privileged access for buses, as is suggested in a scheme now proposed for Southampton. When this scheme is introduced, buses will be able to enter the main road at special access points even when other vehicles cannot; once on, they will be guaranteed the same high level of service as the other vehicles which have entered at the controlled points. More modest ways of using traffic lights to help buses are by phasing the lights in their favour, or by giving buses the power to actuate the traffic signals as they approach.

One warmly advocated and much discussed method of control is road pricing.* As usually proposed, this would involve some kind of meter attached to the vehicle which would register the amount of miles performed or time spent on different parts of the road network, for which the driver would be charged directly. The amount of the charge would depend on the level of congestion prevailing, in other words on the amount of delay and obstruction which the person who chooses to use his vehicle at that time imposes on other road users. Road pricing can be made selective by charging different types of vehicle at different rates; for example, buses need not be charged at all, and the charges imposed on other vehicles could presumably be fixed at a level which would limit the volume of other traffic below the point at which it would disrupt the reliable working of the bus system. Road pricing could also be used in a way which would discriminate between users; for example, disabled drivers who were incapable of using public transport could be charged less either by being given a special meter or by being allowed to claim a rebate.

One great attraction of road pricing is that it is a self-select method of control: it is up to each individual driver to decide for himself, in the light of his particular journey requirements, whether it is worth paying the price or not. This advantage is shared by parking control, in so far as that involves charges. It is also shared by some other means of controlling moving vehicles, although the 'cost' which each traveller has to assess takes the form of a time delay rather than a monetary payment. Thus the reason why, as has been seen, traffic does not in fact 'grind to a halt' in cities at the moment is that there are some people who would rather not travel than pay the cost in terms of the delay inherent in congestion. Another method of self-selection was

* The classic reference for this is *Road Pricing: the Economic and Technical Possibilities* (the Smeed Report), HMSO, 1964. A more technical discussion is given in 'Technical Possibility of Special Taxation in Relation to Congestion caused by Private Users', Professor M. E. Beesley, *Second International Symposium on Theory and Practice in Transport Economics*, European Conference of Ministers of Transport, OECD Publications, Paris, 1968.

mentioned incidentally in Chapter 4, where it was suggested that a system could be devised in Oxford such that it would always be possible to get from any part of the city to any other part by car, but for some journeys the car routes would be so much longer than bus routes that only those who had some very good reason to go by car would do so.

Road pricing raises some doubts from the point of view of equity, since it is a method of rationing based on the ability to pay. How serious this objection would be would depend very much on the aims and administration of any particular scheme. If the aim were to safeguard the operation of the bus system, which is open to all, while at the same time preserving the option of the use of the car, and if the charges were not too high, it should not be too objectionable; even though some of those who chose to use a car would be doing so not because they had any special need to but because they were rich enough to regard the charges as trivial.

Much depends also on how the revenue derived from a charging system would be used.

It is sometimes argued that because motorists have provided the revenue it should be spent in ways which would benefit motorists, but this argument seems to rest on a confusion. The individual motorists who make payments already obtain a service for their money by having the use of relatively uncongested roads, even though there may be some of them who would prefer congested roads and no payment. There is no reason in equity why they should get their money back, and if they did it would defeat the whole purpose of the pricing system, since few people would be deterred from driving by having to pay out money which they would ultimately recover. If any class of people does have a special claim to the money it is those people who have been priced off the roads, who may be said to have fore-gone a right that they used to enjoy and who therefore perhaps deserve some compensation. It is hard to know how those people could be identified specifically enough to repay them, and in any case they might have benefited in other ways, for example by the improvement to the bus system. Probably the best way of re-garding the revenue is simply as another source of taxation revenue, which can be used either to reduce other kinds of taxes, not necessarily connected with motoring, or for the general purposes of government.

A disadvantage of a road pricing system based on meters is that it would require a major effort to fit all motor vehicles with the necessary equipment and generally to set up and administer the system. It would be beyond the powers of individual cities or towns to institute it, and it would not really count as a management measure in the rather special sense that the term is being used in this chapter – something that it would be relatively easy to set up, experiment with and, if necessary, abandon or reverse. Substitutes in the form of toll charges or the payment of fees would be much easier to implement and could be introduced town by town. It was also suggested in Chapter 8 that there might be some advantage in London in using taxi fares as a deliberate method of control, since the necessary equipment for doing so already exists. Although London is a very special case, there might be some application for this idea elsewhere. Since many of the benefits hoped for from the more ambitious methods of road pricing might be obtained either by these less sophisticated methods, or by the other kinds of management measures described in this chapter, it would seem sensible to regard the more ambitious schemes as something to be considered only after other management measures have been tried.

Parking control has been much developed and refined since the days when the control of on-street parking was seen primarily as a means of increasing the capacity and the safety of the road system, although that can still be a legitimate and important objective. An obvious use of parking control is to limit the total number of parking spaces available in an area and thus to limit the total number of journeys by car that can be made there. Control can also be exercised over the times of day that the parking places are available, the lengths of time that drivers may stay, and the charges made, which can themselves vary by time of day. These can be important controls in relation to the purpose and timing of car journeys. For example, if parking spaces are available for only a four-hour period at most, this will discourage commuting by car, and hence ease the rush hour problem. A similar effect can be achieved by not making spaces available until after the morning peak is over. Another possibility, which would apply particularly to off-street parking, would be to impose an extra charge on anyone arriving during the morning peak or leaving during the evening peak. This would have the double advantage of discouraging peak-hour car travel and facilitating car use for those who are obliged to work odd hours when public transport might not be able to cater for their needs.

Through such means as residents' parking schemes, car parking control can discriminate further between different types of user. There might be a good reason for wishing to discourage long-distance car journeys to an area while at the same time preserving the freedom of residents to make short local trips: an example was given in Chapter 8, where it was suggested that people coming to Central London should be encouraged to change from cars to public transport as far out as possible. Another example of discrimination by type of user is the rule or informal convention that has grown up in many places whereby doctors are allowed to park where other drivers may not.

The limitation of parking control is that it is not universal. In all towns there are odd bits of land which are privately owned which are used for parking and it would be difficult to prevent this entirely. In addition, through traffic and taxis are not affected. Whether or not this limitation is important depends on the particular circumstances. In many towns through traffic is not important and would not become so, even if road conditions were improved by the control of local or stopping traffic. But there are other places, as was seen in Chapter 8 in relation to Central London, where any relief brought about by parking control might well be swamped by an increase in travel by vehicles not subject to parking control. Even in such places, it would be wrong automatically to draw the conclusion that parking controls have no part to play: they may still be useful if supplemented by other measures which would impinge on the rest of the traffic.

Controls of the various kinds described are not the only means by which a local authority can intervene in the operation of a town's transport system. The other ways are by subsidising services or by directly providing services of a kind which otherwise would not exist. The argument about subsidies to public transport tends to generate more heat than light on both sides. Opponents tend to argue that public transport services should pay their way like any other commercial enterprise; if the customers are not willing to pay for the service then it should not be provided. Certainly, the burden of proof must always be on those who suggest that the costs of providing any goods or services should be borne by people other than the immediate customers or recipients. But in this case, many other people have an interest in preventing the greater use of private transport which may be the alternative to a subsidised public transport system. In the long run, it may be cheaper to subsidise public transport than to provide roads, so that even from the point of view of minimising public expenditure subsidies may be justified. If subsidies are also a way of providing a much more frequent public transport service than could be supported from fares alone, they may lead to less car use and hence to a reduction in accidents and in various kinds of environmental damage. This may also result in a saving in public expenditure, for example on hospitals, but even if it does not, payments to avoid those unfortunate effects are a perfectly legitimate use of public money.

On the other side, too much is often expected from subsidies. If congestion is the obstacle to providing a fast and reliable bus service, it is not likely that subsidies, or even a completely free service, would induce enough other travellers to switch to the bus for the congestion to be eliminated. The more direct controls described in this chapter will still be necessary, and people who shrink from such measures and think that subsidies are a substitute for them are victims of a dangerous kind of wishful thinking.

Whether or not subsidies have a part to play depends very much on the characteristics of the area. In the centres of large towns, there should be enough patronage for a good public transport system – a public transport system protected in some way from congestion – to make subsidies unnecessary.* But in smaller centres, or in less central areas, there may well be a case for providing more frequent services than could be supported from fares revenue alone.

New services may take the form of bus services specially

* Even in these situations, subsidies may have some part to play as a stop-gap measure until more suitable measures of control can be thought out and implemented. In London, for example, the decline in bus service and patronage continues at a frightening pace. Rising fares are only part of the explanation, and physical or other controls which would protect buses from congestion would also save bus costs and increase patronage and would in this way allow bus fares to be reduced. But in the meantime, the only way to hold fares and prevent further deterioration is by subsidies.

designed for central areas, as have been provided in Leeds,[1] or the various kinds of aids to pedestrian movement, such as moving pavements, which are now in use at some stations and airports in different parts of the world. They can also include various kinds of self-drive cars. Small, low-powered electric cars might be the most suitable form of vehicle, but ordinary cars can be used for the purpose: a scheme of this kind is already operating in Montpellier.

The problems of urban transport are usually discussed only in terms of the movement of people, but there may be great scope for rationalising goods movements too, in the sense that given flows of goods could be handled more cheaply and conveniently in terms of vehicles. Such rationalisation could come about by applying the methods of study that have been developed in recent years to design economical distribution systems for large commercial enterprises to the problem of goods distribution within a town. The town itself, or in large cities particular parts of it, would be treated as the unit of organisation. In the extreme case, there would be a town depot or depots where goods coming from outside the town would be dropped and goods leaving the town for outside destinations would be assembled. Deliveries and collections within the town would be made by vehicles belonging to the town, not necessarily in the legal sense, although that might be appropriate, but in the sense that the carriage of goods within the town would be their sole function.

One benefit might be cheaper goods distribution; there would be extra handling and depot costs, but these might be offset by economies of scale arising particularly from better vehicle utilisation. There should be some reduction in the congestion imposed by goods vehicles on other road users, particularly since it should be easier than at present to control the timing of goods vehicle movements so as to avoid the peak periods of personal travel. A saving in accidents could come about both from a reduction in the mileage performed by goods vehicles, and also because it would be possible to enforce particularly high standards of maintenance on vehicles belonging to the specialised system. For the same reasons, there should be considerable environmental gains, which would be very much enhanced if the specialised system could also make use of specialised vehicles, for example vehicles powered by electricity or gas. This becomes a real possibility when vehicles do not have to travel outside a small area and therefore do not require to be designed to cope with a wide range of conditions. Further economies could arise by sharing garages, maintenance facilities and staff with the passenger public transport system. In the long term, the reduction in road mileage performed by goods vehicles could give rise to a saving in road building costs.

Useful benefits could be obtained from less ambitious schemes of this kind. For example, a depot and secondary distribution system could be set up to supply the shops in a pedestrian shopping street or centre. This might be preferable, in some towns at least, to the alternatives favoured at the moment, which are either to allow goods vehicles into such centres only at night, or to provide access at the back of the shops – an arrangement which may cause more demolition and environmental damage than the pedestrian precinct is worth.

This discussion of management measures has been only cursory. The measures that have been discussed could be extended and refined, and there may be many other possibilities. Even so, enough has been said to indicate the great richness of the means that can be employed to improve the transport system of a town without adding to its road network.

10. The Alternative to a Transportation Study

In Chapter 3, the conventional kind of transportation study criticised, primarily on the grounds that it begged all the main questions, both about the important problems are and about what action the local authority or the government might take in order to solve them. The aim of this chapter is to give a brief outline of a more satisfactory method of study. Such a method must have the following characteristics which the conventional method wholly or partly lacks:

1. It must be concerned with all modes of travel, with pedestrians and cyclists as well as with those who travel by motorised means.
2. It must have regard to the quality as well as the capacity of the transport system.
3. It must take account of present day frustrated travel, of those journeys which people would like to make but do not because they are deterred by the shortcomings of the transport system, as well as of the journeys which are now performed.
4. It must take account of road accidents.
5. It must take account of the way that transport impinges on other things, of the environment as well as of movement.
6. It must provide ways of formulating appropriate proposals as well as testing them.
7. It must allow for all the different kinds of action that might be appropriate to be considered; the introduction of management measures as well as investment in roads and other transport facilities or changes in land use.

The method of study adopted must also, of course, be very much influenced by the circumstances of the particular town in question. The following account is intended to refer to an existing town which is large enough to support its own bus system and for congestion to be a problem; but not so large as to make a railway a serious contender; and which may be growing in accordance with natural population growth, but is not intended to receive extra population from elsewhere. In a town larger than this, even if it were not growing, the number of possible alterations to the transport system that might be suggested is much too large for them all to be considered individually. Some systematic way of grouping and exploring the possibilities must be found. This is a difficult technical problem, which considerably complicates the study outlined below, but raises no new issue of principle. In a town which was expanding, even though it might be a relatively small town, there might be a whole number of land use plans which deserved serious consideration, and this again very much complicates the problem, although the broad approach would remain the same. The only situation that the following account would scarcely fit at all would be a new town, or a town which is expanding so much as to be virtually a new town, but the special problems of new towns need not concern us here.

The first stage of the study of an existing town is to find out how the present situation might be improved – the question that the conventional study implicitly assumes away by setting up the present as an ideal. This stage is likely to involve the same kind of survey of the travel performed in a town as the conventional transportation study, but it would cover travel on foot or by cycle as well as by motorised means, and would be supplemented by surveys on travel conditions. These would investigate the difficulties experienced by travellers by each mode; the mother pushing her pram who finds crossing the road difficult or unsafe, the bus passenger faced with an unreliable service, the motorist or lorry driver who finds he cannot park within striking distance of his destination. The surveys should also examine the causes of these difficulties.

The first stage must also include a careful examination of accident records to identify where, and in what sort of conditions, accidents occur. This leads on naturally to an examination of all the other ways in which travel disturbs other activities, usually summarised under the headings of noise, fumes, visual intrusion and severance.

The question of frustrated travel must also be considered at this stage. To what extent are people failing to make journeys at all, or making them to places or by means different from those that they would prefer, simply because of the shortcomings of the transport system?

All the topics mentioned can and should be studied factually; even, if conditions permit it, frustrated travel. For example, the number of pedestrian or bus journeys made in one part of the town where conditions are poor can be compared with the number made in another part where they are better, or with the number made in other towns where steps have been taken to encourage those modes of travel. But the surveys cannot be purely factual, they must be supplemented by some kind of attitude questions. It is not possible to know what shortcomings in the transport system most deter or concern people without asking them. Of course, the interpretation of such questions poses many problems, as does the interpretation of observed behaviour; and even if it did not, the policy maker cannot be guided blindly by what people tell him. Accidents are a serious matter, whether there is public concern about them or not. There may be special classes of people, for example old people or children, whose problems deserve much more attention than would be suggested by answers to attitude surveys. But although the policy maker has the right to override the findings of any such survey and impose his own views, he should not do so without at least finding out what the public's views are. Still less should he attribute to those for whom he is planning views and aims which they do not possess. It has been seen that in the conventional transportation study there is a constant danger of making wrong inferences about travellers' wishes. Certain aspects of travel, particularly quality aspects, are ignored altogether, although to travel in reasonable comfort on essential journeys may be much more important to people than to be able to make a few more optional journeys. All travel is treated as being desired by the travellers concerned, even though in the extreme case they may be travelling on journeys which they very much resent having to make, such as journeys to ferry children about. Just as manufacturers of consumer goods use attitude surveys and other kinds of direct questioning to ascertain the wishes of their customers, rather than relying on sales figures alone, so should policy makers in this field. The case is indeed far stronger, since there is much more room for the misinterpretation of observed behaviour in the field of urban transport than in most markets.

Urban transport policy takes a long time to formulate and even

longer to implement, especially if new roads or other infrastructure are proposed. Hence, it is not sufficient to rely on an
analysis of present problems; some attempt must be made to
foresee how the situation is likely to develop. Of course, how the
situation will develop in large part depends on what policies are
adopted; it has been seen that the conventional transportation
study implicity assumes, at least when applied to British conditions, that there will be major investment in roads.

The kind of forecast that is required is just the opposite of
this; it should implicitly commit the policy maker as little as
possible. The question to be answered is 'What sort of situation
would we face in the future if we did not intervene at all – that
is if we did nothing except essential maintenance and repairs?'
This is not a question which can be answered with precision,
but in a town where the population was stable it would at least
be a straightforward question. Population growth complicates
the issue, particularly as there may be more than one way of
handling it; the expected growth might all be concentrated in
one part of the town, in which case there might be room for
argument as to which part that should be, or it might be spread
evenly round the town, or some intermediate solution might be
considered. It may be possible to dismiss some of the alternatives
fairly early on, but there may be several serious contenders, each
of which requires to be studied as a separate design. For each
design studied, the best way to deal with population growth is
probably to separate it from the other changes which are likely
to occur; that is, to consider how the town would now be operating if the growth had already taken place. This will involve
amplifying and adjusting the present travel pattern, as established by survey. To adjust the results in order to allow for urban
growth by itself is a relatively straightforward step; to attempt
to take account of urban growth simultaneously with all the
other factors which are going to influence the observed situation
would be both difficult and confusing.

In allowing for the effects of population growth, it may be
necessary to assume some changes in the transport system. For
instance, if the extra population were all to be concentrated in
one part of the town, it would almost certainly have to be accom-
panied by some new road building and would also presumably
warrant some new public transport facilities such as a bus service

to the town centre. These consequential additions to the transport system should be such as to provide the notional residents of the new part of the town with the same level of service as that enjoyed by the actual residents of the existing town.

Having made these adjustments, the next step is to forecast how the situation would develop if a policy of no intervention were adopted. There are three difficulties in being at all precise about this. First, it is extremely difficult to forecast with any precision those characteristics of the population, particularly income and car ownership, which influence travel behaviour. Second, it is hard to know what assumption to make about the effects of changes in those characteristics. It has been seen that it is conventionally assumed that in the future people in a given category, defined principally by income and car ownership, will make the same number of journeys and generally have the same travel behaviour as the people in that category today. This is a doubtful assumption, however, even if the transport system were to expand in a way which would make such regular growth in travel physically possible. Third, and most important, is the question discussed in Chapter 1 of just how travel behaviour adapts to and is affected by the transport system provided. It is not known in detail how this works; at what precise point, for example, congestion is likely to act as a brake on the growth of traffic volumes.

But although precision is not to be hoped for, it should at least be possible to say something about the *direction* in which the situation will develop. If a town is experiencing growing traffic congestion, the downward spiral of reduced public transport usage and service, increasing difficulties for pedestrians and cyclists and general environmental deterioration, one would expect these trends to continue unless something were done to prevent them. The present situation may give some clue as to how far the current trends will go. For example, there may be some main roads on which traffic volumes have settled down to a constant level. But even to know the direction of the trend may be quite sufficient. If one were to try to formulate policy simply on the basis of the present situation, without making any attempt to forecast, one would run the risk of concentrating on problems which might have disappeared, or at least altered substantially, by the time that the policy could be implemented. Forecasts are required to obviate that risk as far as possible. In fact, it would appear that in most towns a forecast of how the situation would develop if there were no intervention would only reinforce the need for the measures which an analysis of the present situation would suggest.

The next step is to formulate policy objectives. This follows fairly naturally from the analysis and forecasts described; the objectives are to remedy the defects that have been discovered in the present or can be foreseen in the future. Some of these defects are more serious than others. By studying the citizens' wishes and adding his own judgement, the policy maker will be able to set his objectives in some kind of order of priority.

Concurrently with the formulation of policy objectives, the policy maker should be examining the constraints that limit the action he may take; particularly the new roads or other structures that he may provide. Some of the constraints will be environmental. The analysis of defects will already have led to the establishment of certain environmental objectives, in terms of setting limits to the amount and type of traffic that should be allowed in certain streets, but environmental considerations will also rule out certain locations for new roads and off-street car parks. Physical constraints in the shape of natural or man-made features must also be taken into account. There are also legal or institutional constraints: the policy maker works within a framework of powers laid down by statute, and clearly there is no point, at least in the short term, in considering courses of action which lie outside those powers. But legal constraints must not be accepted too readily; laws can be changed and it is hardly likely that the legal framework which exists now is the most suitable to deal with the problems that can be foreseen.

Finally, there is the monetary constraint. This is probably the most important of all, both because investment in transport tends to be expensive, and because, given enough money, almost all the other constraints could be overcome or set aside. It is, therefore, essential to establish at the outset how much can be spent on the town, so that time is not wasted in examining infeasible solutions. But it is most important that this sum of money should be treated as a constraint and not as a budget; it represents the maximum that can be spent, but spending up to that maximum still requires justification.

Neither the objectives nor the constraints that have been established by this stage can be treated as absolute requirements. They can only be provisional, since it may turn out that they conflict with each other. But each one stands until it has been shown that there is some such conflict. It may be obvious at that point which of the conflicting specifications is more important and should therefore be retained and which should be dropped or modified. If it is not obvious, then different designs will have to be worked out in detail, one conforming with one subset of the original specifications and the other with another subset, so that a choice can finally be made between two or more complete designs.

This, however, is to anticipate the next stage. Having set out the various specifications (objectives and constraints), the next step is to try to formulate proposals which conform with them. To do so requires an imaginative leap. Proposals will not emerge simply from a process of defining and specifying, however far that process is taken. Equally, of course, good proposals will not emerge simply out of the air, without a prior stage of definition. One reason why imagination is required is that there may be several ways to remedy the defects in the transport arrangements that are now causing complaint, and the most obvious way may not be the best. For example, the best way to deal with the complaint of the mother who has to spend time ferrying her children to the swimming pool may not be to improve the bus service but to build a new swimming pool within walking distance. The best way to deal with the complaint of the motorist who finds it difficult to find a parking place when he goes shopping may not be to provide more parking spaces but to provide a better bus service, so that some of the people who compete with him for parking space may be induced to go by bus instead; this solution will be particularly appropriate if at the same time it meets the complaints of those who have no choice but to go by bus. Thus, although the original analysis of defects had to proceed mode by mode, item by item, when it comes to formulating policy the approach must be broader. The same measure may simultaneously serve several purposes, and the best way to deal with a problem may not be to attack it directly, but to so alter other features of the situation that the original problem no longer arises.

F

Apart from the difficulty that has already been mentioned, that some of the original specifications may conflict with each other, the major difficulty at this stage is to know how the proposals will work. As has been seen, prediction is hard enough even when it is assumed that the present situation is desirable and the relationships observed in it should be maintained. The difficulties become more formidable still if that assumption is questioned, if indeed it is accepted that one aim of policy is to alter the way that the system now works. It is extremely hard to know in advance of implementing any measure what its effect will be.

Models of the kind developed in transportation studies help to some extent. Since they are a means of testing the capacity of roads and transport systems to deal with hypothetical loads, they can show what the effect of some proposal might be, and what it could not be. This is valuable since at least it should prevent proposals being put forward which could not possibly achieve the desired end. But such models are not adequate for making predictions of any fineness; only real life trials will do. Since it is not possible to experiment with building roads or off-street car parks, which once built are built, the trials must be concerned with the kind of control or management measures described in Chapter 9, which are relatively cheap to implement, and can be modified or reversed if they do not succeed. Only when the transport system is properly managed should the question of adding to it arise. Indeed, only then will it be possible to see what additions are called for. It has been seen that one of the crucial weaknesses of the conventional transportation study is that it is often quite unclear how the proposals that are tested by it were originally derived. In order for an investment proposal to be a candidate for testing, it must be shown that it is a means of achieving certain desirable aims that cannot be achieved by management measures alone.

Let us assume, then, that the stage has been reached that the transport system has been reformed but still has certain weaknesses which only investment in new infrastructure can remedy. There may be several alternative proposals and the question is which of them should be chosen, or whether it would be better to allow the defect to remain rather than to spend the money, and perhaps incur the damage, necessary to remove it. Assuming that by now the consequences of adopting each proposal can be predicted with fair confidence, it becomes a matter of how to evaluate those consequences.

It is desirable to conduct the evaluation as far as possible in terms of a common unit; money is the most convenient. But there are many effects which cannot be expressed in monetary terms, at least at present, and perhaps not even in principle.* Ultimately, therefore, the decision remains a matter of judgement. But if the systematic process which has been described is followed, at least the judgement to be made is clear and well defined and the aims which it is hoped to achieve are sensible ones. To reach that position is a great advance. It has been seen that for years the problem at Oxford was wrongly defined as a choice between preserving the High Street or preserving Christ Church Meadow; no such choice need arise, and what the real choice to be made is has still not been made clear. Perhaps it is a choice between the preservation of the entire city and the convenience of travellers, but the terms of that choice have not been defined. Similarly, in London attention has been focused on one particular problem, how to allow the future car owning population to make social trips by car, which is far from being the most urgent transport problem which confronts the city.

There is nothing original in the approach just set out.† It closely resembles the kind of approach developed by town planners long before the advent of transportation studies and is akin also to methods used in other fields of management and investment; all these methods are in the end no more than organised commonsense. But it may appear novel at first sight because of the attitude to road building that has been expressed. It has been suggested that except where new roads are obviously

---

* It is beyond the scope of this book to describe the techniques that have been developed to make an economic appraisal of transport investments, and in particular to put monetary values on benefits, such as time savings, or on losses, such as the noise affecting local residents. For a clear introduction to the subject, with several examples taken from the field of transport, the reader is referred to *Values for Money* by Michael J. Frost, The Gower Press, 1971.

† I have drawn particularly on Chapter One of *Motorways in London*, which any reader interested either in London's problems or in the more general problem of setting about the study of a town will certainly find valuable.

required to support the physical expansion of a town, road building should be one of the last steps to be considered. The reason is that to build roads is expensive, irreversible and very often damaging environmentally; the traffic effect is uncertain and the benefits new roads bring may be obtainable by other simpler means. There should be nothing very startling about this reasoning; no businessman would consider a very expensive addition to his factory if there were good grounds for supposing that the production of the present factory was much less than it should be owing to poor management. But it appears startling because of the deeply ingrained traditional attitudes that have been described, based on the idea that traffic is naturally growing, that this growth should be accommodated and that to fail to provide the necessary roads would have the most grave, if ill-defined, consequences.

It has been seen that even in Central London, where no conceivable amount of road building could cure congestion, the idea that road building must be the main element in any solution was so firmly held that a purely arbitrary target of providing 60% extra capacity was provided, without any regard to the costs and problems that might be entailed. The same instinctive attitude can be seen in other towns. For example, in a letter written by the Welsh Office to a Cardiff MP who was concerned about the proposals to build motorways in Cardiff it was said:

I do not think that anyone can seriously doubt that considerable investment in new roads is going to be necessary in Cardiff in the 1970's. Whatever restraints are practicable on the growing use of private cars in the City and whatever steps are taken to make increasing use of public transport, Cardiff can develop as an administrative and commerical centre and as the Capital of Wales only if vehicles of all kinds can move around and within the City in reasonable comfort. If the Hook Road is not built, the Council will have to look for an alternative and this alternative cannot be of significantly lower capacity than the Hook Road or cost much less. The rate at which funds can be made available will affect the time over which construction will be extended but is unlikely to affect the nature of the road schemes themselves.[1]

It is true that Cardiff is expected to grow, through migration as well as natural increase, and to that extent some new roads are certain to be required. But that has little or nothing to do with the proposed Hook Road, which is a motorway leading to the centre and is intended to form part of a much larger motorway network focused on the centre. The need for such roads should indeed be seriously questioned, and there are at present no grounds for saying that the successful development of Cardiff depends on them, no matter what management measures might be adopted.

Another more general example comes from a letter from the Ministry of Transport to town clerks dated 16 October 1969. The irony of this is that the point of the letter is to urge local authorities to adopt bus priority schemes; the letter is, indeed, something of a historical landmark in its recognition that policy must start with travellers rather than vehicles. Nevertheless, the first paragraph of the letter reads as follows:

As we all know, urban traffic congestion is one of the most intractable problems of the motor age. Yet somehow we have to deal with it if life in our major cities and towns is not to come to a standstill or to become intolerable. Better urban roads are needed and they are being provided. But they are certainly not the whole answer; even if there were no financial constraints there is a limit to what can be done by way of new construction if the character of towns is not to be destroyed.

The danger of life coming to a standstill is illusory and there is no general case for saying that better urban roads are needed. They may be needed in particular towns, but this can be seen only after management measures, such as those which the letter rightly goes on to advocate, have been implemented and their effect assessed.

11. Institutions and Technology

If we are to solve the problems associated with movement in towns, the first step must be to get the theory right. The argument of this book has been that we have been trying to grapple with the problems within a theoretical framework which is inappropriate and too narrow; hence we have ruled out even from consideration lines of policy which must be adopted if adequate solutions are to be found. We have been playing the game by the traditional rules without realising that in modern conditions those rules make it impossible for us to win. The book has tried to show that when the rules are changed all sorts of promising and winning moves become possible.

But if a more appropriate theory is the first requirement, it would be naive to suppose that it would be sufficient. There have been institutional as well as intellectual reasons for our failure to deal with the problems more adequately, and certainly institutional changes will be necessary if the kinds of policy sketched out in this book are to be implemented.

One of the institutional problems is that the skills that are required do not correspond with those that the professions traditionally concerned have provided. The questions that have been discussed in this book all fall within the domain of the town planner, certainly according to the definition of the town planner's function put forward by the then president of the Town Planning Institute in a forceful speech some years ago. 'As I see it, we are concerned to design a rational system of land use and communication routes as a framework into which development can be satisfactorily fitted as regards kind, quantity and timing.'[1]

Transportation studies are ostensibly concerned with very similar problems. It has been argued that in their usual form transportation studies are quite inappropriate to British problems, but the fact that they were nevertheless imported and within a very short time were accepted as a standard procedure is an indication of the weakness of the native tradition. Town planners did not usually possess the basic skills that are required for carrying out such a study: skills in surveys, the formal representation of transport networks and movement, or mathematical modelling more generally. Hence they were in no position to identify and criticise the policy assumptions implicit in the studies that were adopted, still less to put forward constructive alternative suggestions.

Another central argument that has been put forward in this book is that it is no longer possible to separate the questions of what transport facilities should be provided and how those facilities are to be used. This means that the town planner should have at least some familiarity with a whole new range of skills, such as those involved in designing bus routes, or in setting up the kind of municipal distribution system for goods that was briefly described in Chapter 9. But such skills are not usually taught as part of his training.

The other people who have been most concerned with the transport aspects of town planning have been borough engineers. By their training, borough engineers are more at home with problems of road construction and maintenance, or with traffic management in the narrow sense of coping in a safe and orderly way with as large a number of vehicles as possible, than they are with the very different problems of transport operation and management.

Thus, in spite of the enlightened views of many individual members of both the town planning and the engineering professions, there have been institutional pressures as well as theoretical reasons to concentrate attention on solutions which rely heavily on investment and construction rather than management.

Urban transport provision, road building in particular, is expensive, and some systematic method is required of comparing investments and considering what sums of money can be justified in relation to the benefits obtained. A traditional weakness in both professions is that their members are not taught to think in these terms. Town planners have traditionally thought in terms of defining and applying suitable standards, covering all sorts of matters such as daylighting, residential densities, areas of open space required and so on, as well as matters more directly concerned with transport such as car parking. Such standards are an appropriate way of dealing with the very basic matters of public health and safety with which town planning was originally concerned, but it is dangerous to attempt to apply them outside these fields. They can easily lead to spending much more money on obtaining some benefit than it is worth, and they can even create difficulties. For example, it was seen in the discussion on Covent Garden in Chapter 6 that the parking standards adopted bore little relation to the area's transport needs and only compounded the difficulty of finding satisfactory solutions to other problems.

Engineers, too, have traditionally been more concerned with finding the most satisfactory way of accomplishing a stated task, such as building a road or a bridge, than in justifying the need for it. All this helps to explain why the principle 'all traffic demands should be met' has continued to influence policy long after it has become apparent that the costs and damage involved in following it might be very considerable. These weaknesses can only be finally removed by education,* but in the meantime much can be done by involving people with other skills more deeply in planning, for example, economists, systems analysts and people experienced in public transport operation.

Some adjustment is also desirable in the relationship between central and local government, particularly financially. Until very recently, the central government would pay 75% of the costs of approved road building undertaken by local authorities, but would not contribute to alternative kinds of investment. Clearly such an arrangement distorts the decision which a local authority has to make: it gives a great inducement to think in terms of solutions which depend heavily on road building. In recognition of this, recent legislation has made provision for a similar 75% grant to be made towards investment in public transport infrastructure, and it is also now possible for bus undertakings to receive financial help with the cost of new vehicles. Although these steps are in the right direction, they are not such big steps as they may at first appear, since most of the cost of supplying a bus service is in labour rather than equipment. The decisions which a local authority has to make are still

* An interesting analysis of the skills and training required is given in *Transport Planning: the Men for the Job*, a report to the Minister of Transport by Lady Sharp, HMSO, January 1970.

extremely distorted by the financial arrangements. It has been seen in this book that alternatives to building roads in towns may include giving operating subsidies to public transport, or making a change in the disposition of land uses, or instituting some kind of municipal goods delivery system. Institutional and financial arrangements should be such that all those courses of action can be treated as the alternatives that in fact they are, without weighting the terms of choice from the outset in favour of any particular one of them.

Presumably it would be possible to arrange for the central government to give financial help to local authorities in ways which would not distort their decisions, but the more fundamental question that arises is why this sort of relationship between central and local government should exist at all. The beneficiaries of transport improvements in a town are the town's citizens themselves; why should the national taxpayer finance them? The disadvantage of the present arrangements is not merely that it distorts the decision in the way described, but that it very much reduces everybody's incentive to look for inexpensive solutions. The local planner or engineer is likely to be much less averse to spending money provided by the central government than money provided by his local ratepayers; his skill in squeezing money from the government may even be a feather in his cap. Civil servants in central government have a concern to see that public money is properly accounted for and that the correct procedures (including the soothing ritual of a transportation study) are complied with before it is spent. But this is something less than a positive incentive to look for economical solutions, indeed it is not part of the function of a civil servant in central government to look for solutions to local problems at all, except in so far as he may suggest modification to plans submitted to him for approval.

Departmental ministers have no incentive to reduce the spending of their own departments; according to Mr Crosland they both are and ought to be judged by their civil servants by their success in winning battles against their cabinet colleagues and the Treasury for money to be spent by their own departments.*

* Interviewer: 'Turning to other constraints – What about the Treasury? How does a Minister get enough resources for his service?'
Mr Crosland: 'By persuading, arguing, cajoling, exploiting his

The only people who at present have a direct interest in seeing that excessive money is not spent on roads or other local development are the Treasury and the other government departments in direct competition for the funds. But they are in no position to scrutinise local plans in detail or to suggest alternatives, and without this, it is difficult to substantiate any case that excessive expenditure may be involved.

One objection that will be made to giving local authorities more power in this or other fields is that they are simply not capable of undertaking the responsibilities: the quality of the decision makers, in particular the elected members, is not high enough. If there is force in this criticism, and it certainly does not apply universally, one must ask what the causes of this sad situation are. Perhaps one of the reasons why local government fails to attract representatives of a sufficiently high calibre is just because its powers are so limited; the central government has the final say even over matters of quite small detail. If the job were made more interesting and responsible, it might attract better people.

It is not being suggested that such far-reaching changes in the relationship between central and local government are necessary before substantial improvements in urban transport planning can be made. The principle put forward in Chapter 10, that investment in roads or other infrastructure should not be contemplated until management measures have been worked out and implemented, could be incorporated in the conditions to be satisfied before any grants for road building were given by central government. But these procedural reforms are unlikely to be sufficient for the best solutions to be obtained. Local planning, transport and environment are essentially matters for local concern and enthusiasm. The remarkable growth of local societies in the last decade, and the hard work that many of them do, is one of several indications that the concern and enthusiasm exist. At the moment, largely because of the institutional structure, they are shown chiefly in protest and other negative forms of action;

---

political position, being a bloody nuisance in Cabinet . . . And of course it's a primary function of a Minister of any Department to be successful with the Treasury, and his officials will quite rightly judge him by how successful he is.'[2]

institutional reform is required to allow them to find a more positive expression.

If the right institutions are necessary for good solutions to the problems to be found and implemented, technology raises even more important issues, since it could substantially alter the nature of the problems to be solved. The electric car, for example, would eliminate one of the environmental problems now associated with the motor vehicle, that of pollution through fumes. The implementation of some of the measures of control that have been described in this book, for example road pricing, or some of the ways of using traffic lights, also requires technological developments. If all these developments were used to help create transport systems in towns which made people less dependent on their own individual motor vehicles, this could well have an effect on car ownership as well as car use, which would reduce the scale of the problems that face us.

All these are real and substantial contributions, but some people expect much more from technology than that – they look for the development of new forms of transport which would ultimately replace the motor vehicle, at least for most of the uses to which it is now put in towns. There is no reason to think that human ingenuity is now exhausted, or that technology will not provide alternatives which are superior to the motor vehicles for at least some of the purposes for which it is now the most appropriate means of transport.* But the motor vehicle has such a central position in the economy that the transition to anything else could only be very gradual; for that reason alone, it would be wrong to look for technology for any immediate solutions to our problems. It would also be unwise to write off the technological possibilities of the motor vehicle too quickly. As it stands, it is a remarkable and versatile machine and is no doubt capable of much further development. Advocates of technologically new systems tend to argue something like this: 'Our present system is in a mess; our present system is based on the motor vehicle; therefore an alternative to the motor vehicle must be found.' But this reasoning would only hold if we were now making the best possible use of the motor vehicle, and a

* Some existing and proposed unorthodox means of transport are described and illustrated in *New Movement in Cities*, Brian Richards, Studio Vista and Rheinhold Publishing Corporation, 1966.

central argument of this book has been that we are not. New inventions may help even further, but the existing technology offers all that we actually need. The problem is to use what we have well, which is a matter of organisation not technology. In the struggle of towns against traffic, the question is which is to be master – that's all.

References

CHAPTER 1

1 See *Daily Life in Ancient Rome*, J. Carcapino, Penguin Books, 1956, page 57ff.
2 Quoted in 'Applications of New Technology to the Transport of Urban Goods' by C. Beaumont Lewis, proceedings of OECD seminar *The Urban Movement of Goods*, held in Paris, April 1970.
3 *Design and Layout of Roads in Built-up Areas*, HMSO, 1946, para 52.
4 Ibid, paras 58 & 59.
5 Ibid, para 68.
6 Ibid, paras 49, 50, 60–63, 99ff, 104ff.
7 Ibid, paras 21, 178, 179.
8 *Town Planning and Road Traffic*, Sir H. Alker Tripp, Arnold, 1942, page 42.
9 Ibid, page 75.
10 *Design and Layout of Roads in Built-up Areas*, paras 109ff.
11 *Town Planning and Road Traffic*, page 53.
12 *Design and Layout of Roads in Built-up Areas*, para 20.
13 Ibid, para 52.
14 *Town Planning and Road Traffic*, page 59.
15 'Evolving Concept of Transportation Engineering', Norman Kennedy, published in *Traffic Engineering and Control*, July 1963.
16 'Repeat traffic studies in 1967 in eight towns previously surveyed in 1963/4', Road Research Laboratory Report LR 390.
17 *The Death and Life of Great American Cities*, Jane Jacobs, Pelican Books, 1964, pages 374ff.
18 'Foot Streets in Four Cities', Norwich City Council, November 1966, para 5.02.
19 *Motorways in London*, J. Michael Thomson *et al*, Duckworth, 1969, Chapter 2, Section 1.
20 *Roads in Urban Areas*, HMSO, 1966, page 2.

21 'The planning of ring roads with special reference to London',
Excerpts from the Proceedings of the Institition of Civil Engineers, Part II, Vol. 5, June 1956, page 127.

CHAPTER 2

1 Speech by Lord Montagu, quoted in *The Motor*, 29 March 1927.
The immediate source for this is *The Motor Car and Politics
1896–1970*, William Plowden, The Bodley Head, 1971, page 401.
Mr Plowden's book contains many references and quotations
indicative of the kind of attitude discussed in this chapter.

2 *Traffic in Towns*, Report of the Steering Group chaired by Sir
Geoffrey Crowther, HMSO, 1963, para 30.

3 Speech by Mr Edward Heath, reported in *Motor Industry Bul-
letin*, Society of Motor Manufacturers and Traders, 27 November
1966. The immediate source is *The Motor Car and Politics
1896–1970*, page 386.

4 *Town Planning*, Thomas Sharp, Penguin Books, 1942.

5 The survey evidence is reviewed in *Transport Strategy in London*,
London Motorway Action Group and London Amenity and
Transport Association, 1971, Appendix Y.

CHAPTER 3

1 *Town Planning and Road Traffic*, Sir H. Alker Tripp, Arnold,
1942, Chapter 6.

2 *Traffic in Towns*, HMSO, 1963, paras 90, 442.

3 Ibid, paras 108–110.

4 *Town Planning and Road Traffic*, pages 54–57.

5 *Traffic in Towns*, paras 104–107.

6 *Town Planning and Road Traffic*, page 68.

7 *Traffic in Towns*, paras 118–120.

8 Ibid, para 102.

9 *Town Planning and Road Traffic*, page 61.

10 *Traffic in Towns*, para 116.

11 Ibid, paras 74, 449–456.

12 Ibid, para 108, Figure 59.

13 Ibid, para 113.

14 Ibid, para 114.

15 Ibid, Figure 99, paras 156 and 157.

16 Ibid, paras 97, 116, 173, 289, 366; Appendix 2: paras 16–21.

17 *Traffic Engineering and Control*, Vol 6, No 1, May 1964, pages 39ff.

18 *Traffic in Towns*, Figure 162.

19 Ibid, para 444.

20 *Town Planning and Road Traffic*, pages 42, 77.

21 *Traffic in Towns*, para 108.

22 Ibid, para 110.

23 *Introduction to Transportation Planning*, M. J. Bruton, Hutchinson, 1970.

24 *Motorways in London*, J. M. Thomson *et al*, Duckworth, 1969, page 27.

CHAPTER 4

1 Quoted in *Oxford Central Area Study*, Scott Wilson Kirkpatrick and Partners and Hugh Wilson and Lewis Womersley, December 1968, para 1.11.

2 *Oxford Traffic Survey 1957*, University Press, Oxford, 1959.

3 *Report of the Oxford Roads Enquiry*, Ministry of Housing and Local Government, March 1961, paras 279–284

4 Ibid, paras 46, 60.

5 Ibid, paras 51, 56.

6 Transcript of 1960 Inquiry, Day 1, page 18.

7 *Report of the Oxford Roads Enquiry*, para 57.

8 Ibid, para 300.

9 Ibid, para 272.

10 *Traffic in Towns*, HMSO 1963, para 391.

11 'Proposals for Alterations or Additions to the City of Oxford Development Plan, Inquiry into Objections and Representations', Department of the Environment, May 1971, paras 235, 237.

12 *Oxford Central Area Study*, Table 32. Design G in this table corresponds in essentials to the plan approved by the Minister.

13 Ibid, para 8.40.

14 Ibid, paras 5.4, 5.15ff.

15 Mr Munby's views were reported at some length in the *Oxford Mail* of 19 August 1964.

16 *Oxford Central Area Study*, Table 15, para 9.11.

17 Ibid, paras 2.15, 12.4.

18 Ibid, para 2.3.

19 Ibid, para 5.25. See also last sentence of para 2.5.

20 Ibid, para 2.5.

21 Ibid, paras 9.7, 9.8, 9.27 etc.

22 'Proposals for Alterations or Additions to the City of Oxford Development Plan, Inquiry into Objections and Representations', paras 87 and 88.
23 Ibid, para 439.

CHAPTER 5

1 *London Road Plans 1900–1970*, C. M. Buchanan, GLC Intelligence Unit Research Report No 11, GLC, 1971.
2 *Motorways in London*, J. M. Thomson *et al*, Duckworth, 1969.
3 *Transport Strategy in London*, London Motorway Action Group and London Amenity and Transport Association, 1971.
4 *The County of London Plan*, J. H. Forshaw and P. Abercrombie, prepared for the LCC, published by Macmillan, 1943, para 193.
5 *Greater London Plan 1944*, P. Abercrombie, HMSO, 1945, para 154.
6 *County of London Plan*, para 207.
7 Ibid, para 212.
8 Ibid, paras 199 and 202.
9 Ibid, para 11.
10 Ibid, para 69.
11 *Greater London Plan*, para 36.
12 *County of London Plan*, paras 29 to 32.
13 Ibid, para 11.
14 *Greater London Plan*, paras 11, 197.
15 *County of London Plan*, para 244.
16 *Greater London Plan*, para 198.
17 Ibid, paras 33, 71; *County of London Plan*, paras 57, 585–587.
18 *Proceedings of the Institution of Civil Engineers*, Part II, Vol 5, page 106.
19 *County of London Plan*, paras 74, 235.
20 *Greater London Plan*, paras 33, 177(1).
21 Ibid, para 157.
22 *County of London Plan*, para 240.
23 Ibid, paras 30, 207.
24 Ibid, para 221.
25 *Proceedings of the Institution of Civil Engineers*, Part II, Vol. 5, page 107.
26 *North East London, A report to the Greater London Council*, Colin Buchanan and Partners, June 1970.
27 Greater London Development Plan Inquiry Proof E 12/5, submitted by the Department of the Environment, December 1970.

CHAPTER 6

1 *Roads in England and Wales*, Report by the Minister of Transport for the Year ended 31 March 1964, HMSO 1964, paras 149, 150.
2 'The Value of Traffic Management', J. M. Thomson, *Journal of Transport Economics and Policy*, Vol 2, No 1, January 1968.
3 *Piccadilly Circus*, HMSO, 1965, para 10.
4 Ibid, para 47.
5 Ibid, para 10.
6 Ibid, Appendix 1.
7 Ibid, paras (13), 32–41.
8 Ibid, paras (8), (9), Chapters 3 and 4.
9 Ibid, para 60.
10 Ibid, para 131.
11 *Traffic in Towns*, HMSO, 1963, Figure 186.
12 *Piccadilly Circus*, paras 38–40.
13 Ibid, paras 61, 122, 136, 137.
14 Ibid, para 45.
15 Ibid, Appendix 2, Figure 5.
16 Ibid, paras 45, 75.
17 Ibid, para 79.
18 *Piccadilly Circus*, paras 49–55, 64–66.
19 Ibid, para 62.
20 Ibid, paras 63, 70–72.
21 *Covent Garden's Moving*, Greater London Council, 1968, para 1.
22 Ibid, para 9.
23 Ibid, para 11.
24 Ibid, para 12.
25 Ibid, para 256.
26 *Covent Garden, the Next Step*, Greater London Council, 1971, paras 42–46 and Figures 2 & 3.
27 Ibid, para 13, as amplified by Inquiry evidence.
28 *Covent Garden, The Next Step*, Appendix C.
29 *Covent Garden's Moving*, para 96; see also evidence prepared by the Covent Garden Community.
30 *Covent Garden's Moving*, paras 154–157.
31 Ibid, Appendix I.
32 Ibid, Table 43.
33 Ibid, Appendices E & G.
34 Ibid, Appendix L.
35 Ibid, paras 133, 163, C29; *Covent Garden, The Next Step*, para 13.
36 *Covent Garden's Moving*, para 336.

37 *Sunday Telegraph*, 4 July 1971.
38 *Covent Garden's Moving*, para 248.
39 Ibid, para 274; *Covent Garden, The Next Step*, para B6.
40 *Covent Garden's Moving*, paras 331, P3, P11.
41 Ibid, paras 215, 219, 304, E12.
42 Covent Garden CDA Public Inquiry, 1971. Proof of evidence submitted by Mr Panther of the Department of Planning and Transportation of the GLC, para 3.
43 Ibid, Appendix 2; *Covent Garden, The Next Step*, Appendix B, para 6.
44 *Covent Garden's Moving*, para G4.
45 Ibid, paras 6, 39–42, 49–53, 89.
46 Ibid, para 39.
47 Ibid, para 205.
48 *Greater London Development Plan Written Statement*, Greater London Council, 1969, paras 7.17 and 7.18.
49 GLDP Inquiry Proof E11/1, para 8.33.
50 GLDP Inquiry Proof E15/1, para 16.
51 GLDP Inquiry Proof E12/1, paras 7.5. 4—7. 5.9.
52 GLC Background paper B476, *Vehicle Access to Central London*, GLC, June 1971, Appendix B, para 3.6.
53 GLDP Inquiry Proof E12/1, Table 8.1.

CHAPTER 7

1 GLC minutes, 6 April 1969, pages 347ff.
2 Joint report of the Planning and Communications Committee and the Highways and Traffic Committee of the GLC, submitted to the Council 19 July 1966.
3 *Transport Strategy in London*, Section 3.3.
4 *London Transportation Survey, Part 3*, Freeman Fox, Wilbur Smith and Associates for the GLC, 1968.
5 *Report of Studies*, GLC 1969, para 6.253.
6 GLC Research Memorandum 159, July 1969, Tables A1 and A2.
7 *Report of Studies*, para 6.173 and Figure 6.42; see also 'Transportation Studies: A review of results to date from typical areas, 1. London' article by B. V. Martin, then Chief Engineer in the Department of Highways and Transportation, GLC, para 20. This article is published in *Proceedings of the Transportation Engineering Conference* held in London 23–26 April 1968, Institution of Civil Engineers, 1968.

8 Joint report of the Highways and Traffic Committee and the Planning and Communications Committee of the GLC, GLC minutes 21 November 1967, pages 694ff.

9 Joint report of the Policy Steering Committee, the Planning and Transportation Committee and the Strategic Planning Committee, GLC minutes, 27 January 1970, pages 42ff.

10 GLDP Inquiry Proof E12/1, Table 6.5, para 7.5.3.

11 GLDP Inquiry Transcript, Day 75, pages 9 and 10.

12 *Transport Strategy in London*, Sections 4, 5, 6.

13 GLDP Inquiry Proof E12/1, Table 10.2.

14 GLDP Inquiry Proof E11/1, para 8.30.

15 GLDP Inquiry Proof E12/1, para 2.10.9.

16 Ibid, para 2.13.4.

17 'The London Transportation Study and Beyond', T. M. Ridley and J. O. Tresidder, *Regional Studies*, Vol 4, Pergammon Press, 1970.

18 *The Times*, 7 January 1969.

19 GLDP Written Statement, paras 2.9, 2.10.

20 *A reply to 'Motorways in London', the report by J. Michael Thomson*, British Roads Federation, 1969, page 16.

21 *Transport Strategy in London*, Table 4.1, para 7.5.15.

22 GLDP Inquiry Proof E12/1, para 3.7.2.

23 Ibid, para 5.9.2.

24 GLDP Written Statement, para 5.9.

25 *Report of Studies*, para 6.255.

26 GLDP Written Statement, para 5.5; *Report of Studies*, para 6.223.

27 *Report of Studies*, para 6.255.

28 GLDP Inquiry Proof E12/1, paras 7.6.2, 8.2.1, 11.4.1.

29 *North East London*, para 199.

30 GLDP Inquiry Proof E12/1, paras 1.2.7, 1.2.8, 3.6.2, 6.2.11.

31 Ibid, para 6.2.5.

32 GLDP Written Statement, para 5.9.

33 GLDP Inquiry Proof E12/1, para 2.2.4.

34 Ibid, para 2.13.1.

35 Ibid, para 3.6.6.

36 Ibid, paras 3.6.10, 6.5.2.

37 *Transport in London: A balanced policy*, GLC, March 1970, page 6.

38 *Towards a Greater London*, Conservative Central Office, 1970.

39 *A Secondary Roads Policy*, GLC, 1970, page 3.

40 *London Transportation Study, Part 3*, para 22.57.

41 *Report of Studies*, para 6.2.69.

42 *London Transportation Study, Part 3*, paras 228, 229.

43 *Report of Studies*, para 122.
44 *A Secondary Roads Policy*, page 3.

CHAPTER 8

1 'An evaluation of two proposals for traffic restraint in Central London', J. M. Thomson, *Journal of the Royal Statistical Society*, Series A, Vol 30, Part 3, 1967.
2 See *Report of Studies*, Appendix Table 6.5, as amended and amplified by GLC Research Memorandum 159.
3 GLDP Written Statement, Section 8.

CHAPTER 9

1 *Planning and Transport the Leeds Approach*, HMSO, 1969, page 19.

CHAPTER 10

1 Extract from letter dated 6 March 1969 from the Welsh Office to Mr E. Rowlands, then MP for Cardiff North, quoted in *Joint Report of Chief Officers on part of the Short Term Primary Highway Programme*, City of Cardiff, April 1971, para 2.5.

CHAPTER 11

1 Presidential Address of the Town Planning Institute, 1965. Lewis B. Keeble, MC, BSC, MA, FRICS, *TPI Journal*, Vol 51, page 357.
2 'The Art of Wielding Whitehall Power', interview with Mr Anthony Crosland, *The Sunday Times*, 26 September 1971.

Index

household surveys (trip generation rates), 41–5

inter-suburban roads *see* orbital roads
Introduction to Transportation Planning, 41n

Jacobs, Jane, 17
Jones, Inigo, 99

LCC (London County Council), 119; Development Plan (1951) of, 86, 88; *see also* Piccadilly
Leeds, bus services in, 153
local authorities, relations between central government and, 168–70
local roads, segregation of through traffic from, 13, 20–2, 38, 123
London, peak-hour travelling in, 27–8, 89; Abercrombie's plans for, 79–87; motorways, 79, 92, 110–33; Inner, 89, 91, 122, 139–43 Central London: traffic management in, 88–9; Piccadilly Circus, 89–98; Covent Garden, 98–106; present policies, 106–9; alternative strategies, 133–39
London Road Plans, 79n
London Traffic Management Unit, 88–9
London Transport, 135, 136
LTS (London Traffic Survey/London Transportation Study), 86–7, 91, 95, 97, 98, 110, 114, 116, 120, 122, 127, 128, 130, 132

The Mall, Sunday closure of, 146
Ministry of Housing and Local Government *see* Piccadilly Circus
Ministry of Transport, 1964 Annual Report, 88–9; *see also Design and Layout of Roads in Built-up Areas;* Piccadilly Circus; *Roads in Urban Areas; Traffic in Towns*

modal split *see* competing modes of travel
motorways, 39, 40; London, 79, 85, 87, 110–32; Cardiff proposal, 164–5
Motorways in London, 18n, 79n, 163n
moving pavements, travelators, 136, 153
multi-storey car parks, 69, 70, 145; *see also* parking

New Movements in Cities, 171n
Norwich, Buchanan scheme for, 38n
no waiting restrictions, 94

optional traffic, Buchanan's definition of, 34, 37
orbital (or tangential) roads, 87, 111, 112–14, 116, 125–7
Oxford redevelopment, 58–78, 163
Oxford Replanned, 58n

parking, parking control, 15, 17, 21, 32, 145, 148, 150–1; in Oxford, 68, 69, 70, 72, 73, 78; in London, 88, 94, 99, 100, 101, 108, 109, 135, 138, 142–3
pedestrians, 25, 26, 57, 136, 142, 153, 155, 156; precincts for, 12, 13, 20, 22, 33, 68, 80, 154; Buchanan's view of, 35 &n; in transportation studies, 52; Abercrombie's view, 80–1; in Oxford, 68, 73, 74; Piccadilly upper level scheme for, 93, 94, 95, 96; in Covent Garden area, 99, 100, 101, 105, 106
Piccadilly Circus, 79, 89–98, 108
'pinch points', 93
Planning and Communications Committee (GLC), 110–12
population growth, 158–59
Port of London, removal to Tilbury of, 86
prediction *see* transportation studies

public transport, 12, 18, 25, 26, 27–8, 38–9, 55, 57, 119, 146, 147–8; transportation studies' bias against, 52–3; in Oxford, 73–4, 76–7; in London, 80, 96–7, 99, 107, 108, 116–17, 119, 120, 128–29, 135–6, 137, 142; subsidies for, 145, 151–2, 168, 169

railways and underground, 82, 87, 107, 119, 135
Regional Studies, 120–1
Report of Studies (GLC), 123, 130
Report of the Oxford Road Enquiry, 58n
Richards, Brian, 171n
ring/radial road system, 13–14, 19, 21, 79, 85, 83–7, 110, 111, 125–6
road management *see* traffic control
road pricing, 21, 56, 138, 148–50, 171
Road Research Laboratory, 61, 65, 89
road safety/accidents, 14, 16, 21, 155, 156, 157
Roads in Urban Areas, 20

segregation (of traffic), 12–13, 14, 20–2; vertical, 32–3, 35, 94, 95, 100
self-drive cars, 153
Sharp, Lady, 168n
Sharp, Thomas, 29, 58n
speed limits, 146
sub-arterial roads, 13, 31, 39, 40

taxis, 36, 39, 96, 133–4, 137, 138, 146, 150
Thomson, J. M., 138n
through traffic, 20–2, 38, 80, 99, 103, 104, 123, 137

Town Planning and Road Traffic, 13–14
Traffic and Engineering Control, 28
traffic assignment, 49–50, 54, 67, 70
traffic control/management, traditional approach, 15, 21–2; the case for controls, 23–30; Buchanan's view, 33–4, 38, 40; in Oxford, 68, 73–8; in London, 88–9, 93–4, 108–9, 136–39, 142–3; management measures discussed, 143–54
Traffic in Towns (Buchanan Report), 20, 31–41, 56, 66, 94, 106
traffic jams, 17n, 26
traffic lights, 93, 136, 147–8, 171
Transport (London) Act (1969), 119
Transport Planning: the Men for the Job, 168n
Transport Strategy in London, 79n
transportation studies, 41–56; the alternative to, 155–65; skills and education required for, 166–168; *see also* LTS
trend prediction, Oxford, 75–6
trip distribution, definition of, 46–7, 54
trip generation, 41–6, 51–2, 54
Tripp, Sir Alker, 13–14, 15, 20, 21–2, 31, 32, 33, 39, 80, 83, 125, 145

United States, traffic growth in, 16; transportation studies in, 41, 51, 52

Values for Money, 163n

Walker, P. (Secretary of State), 70
Washington Square Park (New York) case, 17
Westminster City Council, 91